NO MAN IS AN ISLAND

Other books by Thomas Merton
published by Burns & Oates

THE ASCENT TO TRUTH
BREAD IN THE WILDERNESS
CONJECTURES OF A GUILTY BYSTANDER
THE LIVING BREAD
THE NEW MAN
RAIDS ON THE UNSPEAKABLE
SOLITUDE AND LOVE OF THE WORLD
THOUGHTS IN SOLITUDE
THOUGHTS ON THE EAST
THE WAY OF CHUANG-TZU
THE WISDOM OF THE DESERT

Also published by Burns & Oates

John Howard Griffin
THOMAS MERTON: THE HERMITAGE YEARS

NO MAN IS AN ISLAND

THOMAS MERTON

BURNS & OATES
A Continuum imprint
LONDON • NEW YORK

First published in Great Britain in 1955
BURNS & OATES
A Continuum imprint
The Tower Building, 11 York Road, London SE1 7NX
370 Lexington Avenue, New York, NY 10017-6503
www.continuumbooks.com

Eleventh impression 1997

Reprinted 2001 (twice)

ISBN 0 86012 004 X

Printed and bound in Great Britain
by Biddles Ltd, *www.biddles.co.uk*

DILECTISSIMIS FRATRIBUS
SCHOLASTICIS ET NEO-SACERDOTIBUS
IN ABBATIA B.V.M. de GETHSEMANI
AUCTOR, QUI ET MAGISTER SPIRITUS,
PERAMANTER
D. D.

Contents

Author's Note

LEAVING system to others, and renouncing the attempt to lay down universal principles which have been exposed by better men elsewhere,* I only desire in this book to share with the reader my own reflections on certain aspects of the spiritual life. I consider that the spiritual life is the life of man's real self, the life of that interior self whose flame is so often allowed to be smothered under the ashes of anxiety and futile concern. The spiritual life is oriented toward God, rather than toward the immediate satisfaction of the material needs of life, but it is not, for all that, a life of unreality or a life of dreams. On the contrary, without a life of the spirit, our whole existence becomes unsubstantial and illusory. The life of the spirit, by integrating us in the real order established by God, puts us in the fullest possible contact with reality—not as we imagine it, but as it really is. It does so by making us aware of our own real selves, and placing them in the presence of God.

This book is, then, a sequel to a previous volume called *Seeds of Contemplation*. But instead of going on from where that book left off, it goes back to cover some of the ground that was taken for granted before the earlier volume began. This book is intended to be simpler, more fundamental, and more detailed. It treats of some of the

* For example, *The Love of God* by Dom Aelred Graham, O.S.B., London, 1939; *Transformation in Christ* by Dietrich von Hildebrand, New York, 1948; *The Theology of the Spiritual Life* by J. de Guibert, S.J., London, 1954.

basic verities on which the spiritual life depends. It is dedicated to the scholastics studying for the priesthood at the Abbey of Gethsemani, who will perhaps recognize in it some notions they have received in spiritual direction.

Fr. M. Louis, O.C.S.O.

Abbey of Gethsemani
January, 1955

Prologue

No matter how ruined man and his world may seem to be, and no matter how terrible man's despair may become, as long as he continues to be a man his very humanity continues to tell him that life has a meaning. That, indeed, is one reason why man tends to rebel against himself. If he could without effort see what the meaning of life is, and if he could fulfil his ultimate purpose without trouble, he would never question the fact that life is well worth living. Or if he saw at once that life had no purpose and no meaning, the question would never arise. In either case, man would not be capable of finding himself so much of a problem.

Our life, as individual persons and as members of a perplexed and struggling race, provokes us with the evidence that it must have meaning. Part of the meaning still escapes us. Yet our purpose in life is to discover this meaning, and live according to it. We have, therefore, something to live for. The process of living, of growing up, and becoming a person, is precisely the gradually increasing awareness of what that something is. This is a difficult task, for many reasons.

First of all, although men have a common destiny, each individual also has to work out his own personal salvation for himself in fear and trembling. We can help one another to find out the meaning of life, no doubt. But in the last analysis the individual person is responsible for living his own life and for "finding himself." If he persists in shifting this responsibility to somebody else, he fails to find out the meaning of his own existence. You cannot tell me who I am, and I cannot tell you who you are. If you do not know your own identity, who is going to identify you? Others can give you a name or a number, but they can never tell you who you really are. That is something you yourself can only discover from within.

That brings us to a second problem. Although in the end we alone are capable of experiencing who we are, we are instinctively gifted in watching how others experience themselves. We learn to live by living together with others, and by living like them—a process which has disadvantages as well as blessings.

The greatest of disadvantages is that we are too prone to welcome everybody else's wrong solution to the problems of life. There is a natural laziness that moves us to accept the easiest solutions—the ones that have common currency among our friends. That is why an optimistic view of life is not necessarily always a virtuous thing. In a time like ours, only the coarse grained still have enough resistance to preserve their fair-weather principles unclouded by anxiety. Such optimism may be comfortable: but is it safe? In a world where every lie has currency, is not anxiety the more real and the more human reaction?

Now anxiety is the mark of spiritual insecurity. It is the fruit of unanswered questions. But questions cannot go unanswered unless they first be asked. And there is a far worse anxiety, a far worse insecurity, which comes from being afraid to ask the right questions—because they might turn out to have no answer. One of the moral diseases we communicate to one another in society comes from huddling together in the pale light of an insufficient answer to a question we are afraid to ask.

But there are other diseases also. There is the laziness that pretends to dignify itself by the name of despair, and that teaches us to ignore both the question and the answer. And there is the despair which dresses itself up as science or philosophy and amuses itself with clever answers to clever questions—none of which have anything to do with the problems of life. Finally there is the worst and most insidious despair, which can mask as mysticism or prophecy, and which intones a prophetic answer to a prophetic question. That, I think, is likely to be a monk's professional hazard, so I purify myself of it at the beginning, like Amos who complained, "I am not a prophet, nor am I the son of a prophet, but I am a herdsman, plucking wild figs" (Amos 7 : 14).

The prophetic illusion—which is quite common in our time—is at the opposite extreme from the gregarious illusion, which is more common still in every time. The false prophet will accept any answer, provided that it is his own, provided it is *not* the answer of the herd. The sheep mentality, on the other hand, accepts any answer that circulates in its own flock, provided only that it is *not* the answer of a prophet who has not been dead for at least five hundred years.

If I know anything of intellectual honesty, and I am not so certain that I do, it seems to me that the honest position lies somewhere in between. Therefore the meditations in this book are intended to be at the same time traditional, and modern, and my own. I do not intend to divorce myself at any point from Catholic tradition. But neither do I intend to accept points of that tradition blindly, and without understanding, and without making them really my own. For it seems to me that the first responsibility of a man of faith is to make his faith really part of his own life, not by rationalizing it but by living it.

After all, these meditations are musings upon questions that are, to me, relatively or even absolutely important. They do not always pretend to be final answers to final questions, nor do they even claim to face those questions in the most fundamental possible terms. But at least I can hope they are thoughts that I have honestly thought out for myself and that, for better or for worse, mean something in my own life and in the lives of those I live with. They point, therefore, toward what seems to me to be the meaning of life. They do not aim to include everything that life can possibly mean, nor do they take in a broad general view of all that matters. They are simply observations of a few things that seem to me to matter. If there is a thread of unity running through them all, I should say it was the following idea:

What every man looks for in life is his own salvation and the salvation of the men he lives with. By salvation I mean first of all the full discovery of who he himself really is. Then I mean something of the fulfilment of his own God-given powers, in the love of others and of God. I mean also the discovery that he cannot find himself in himself alone, but that he must find himself in and through others. Ultimately, these propositions are summed up in two lines of the Gospel: "If any man would save his life, he must lose it," and, "Love one another as I have loved you." It is also contained in another saying from St. Paul: "We are all members one of another."

The salvation I speak of is not merely a subjective,

psychological thing—a self-realization in the order of nature. It is an objective and mystical reality—the finding of ourselves in Christ, in the Spirit, or, if you prefer, in the supernatural order. This includes and sublimates and perfects the natural self-realization which it to some extent presupposes, and usually effects, and always transcends. Therefore this discovery of ourselves is always a losing of ourselves—a death and a resurrection. "Your life is hidden with Christ in God." The discovery of ourselves in God, and of God in ourselves, by a charity that also finds all other men in God with ourselves is, therefore, not the discovery of ourselves but of Christ. First of all, it is the realization that "I live, now not I, but Christ liveth in me," and secondly it is the penetration of that tremendous mystery which St. Paul sketched out boldly—and darkly—in his great Epistles: the mystery of the recapitulation, the summing up of all in Christ. It is to see the world in Christ, its beginning and its end. To see all things coming forth from God in the *Logos* who becomes incarnate and descends into the lowest depths of His own creation and gathers all to Himself in order to restore it finally to the Father at the end of time. To find "ourselves," then, is to find not only our poor, limited, perplexed souls, but to find the power of God that raised Christ from the dead and "built us together in Him unto a habitation of God in the Spirit" (Ephesians 2:22).

This discovery of Christ is never genuine if it is nothing but a flight from ourselves. On the contrary, it cannot be an escape. It must be a fulfilment. I cannot discover God in myself and myself in Him unless I have the courage to face myself exactly as I am, with all my limitations, and to accept others as they are, with all *their* limitations. The religious answer is not religious if it is not fully real. Evasion is the answer of superstition.

This matter of "salvation" is, when seen intuitively, a very simple thing. But when we analyse it, it turns into a complex tangle of paradoxes. We become ourselves by dying to ourselves. We gain only what we give up, and if we give up everything we gain everything. We cannot find ourselves within ourselves, but only in others, yet at the same time before we can go out to others we must first find ourselves. We must forget ourselves in order to become truly conscious of who we are. The best way to love ourselves is to love others, yet we cannot love others unless we love ourselves, since it is written, "Thou shalt love thy neighbour as thyself." But if we love ourselves in the wrong way, we become incapable of loving anybody else. And indeed when we love ourselves wrongly we hate ourselves; if we hate ourselves we cannot help hating others. Yet there is a sense in which we must hate others and leave them in order to find God. Jesus said: "If any man come to me and hate not his father and his mother . . . yea and his own life also, he cannot be my disciple" (Luke 14:26). As for this "finding" of God, we cannot even look for Him unless we have already found Him, and we cannot find Him unless He has first found us. We cannot begin to seek Him without a special gift of His grace, yet if we wait for grace to move us, before beginning to seek Him, we will probably never begin.

The only effective answer to the problem of salvation must therefore reach out to embrace both extremes of a contradiction at the same time. Hence that answer must be supernatural. That is why all the answers that are not supernatural are imperfect: for they only embrace one of the contradictory terms, and they can always be denied by the other.

Take the antithesis between love of self and love of another. As long as there is question of material things, the two loves are opposed. The more goods I keep for my own enjoyment, the less there are for others. My pleasures and comforts are, in a certain sense, taken from someone else. And when my pleasures and comforts are inordinate, they are not only taken from another, but they are stolen. I must learn to deprive myself of good things in order to give them to others who have a greater need of them than I. And so I must in a certain sense "hate" myself in order to love others.

Now there is a spiritual selfishness which even poisons the good act of giving to another. Spiritual goods are greater than the material, and it is possible for me to love selfishly in the very act of depriving myself of material things for the benefit of another. If my gift is intended to bind him to me, to put him under an obligation, to exercise a kind of hidden moral tyranny over his soul, then in loving him I am really loving myself. And this is a greater and more insidious selfishness, since it traffics not in flesh and blood but in other persons' souls.

Natural asceticism presents various insufficient answers to this problem. Each answer contains a hidden temptation. The first is temptation to the hedonism of Eros: we deny ourselves just enough to share with one another the pleasures of life. We admit a certain selfishness, and feel that in doing so we are being realistic. Our self-denial is, then, just sufficient to provide us with a healthy increase in our mutual satisfactions. In a bourgeois world, Eros knows how to mask as Christian charity.

Next comes the temptation to destroy ourselves for love of the other. The only value is love of the other. Self-sacrifice is an absolute value in itself. And the desire of the other is also absolute in itself. No matter what the lover desires, we will give up our life or even our soul to please him. This is the asceticism of Eros, which makes it a point of honour to follow the beloved even into hell. For what greater sacrifice could man offer on the altar of love than the sacrifice of his own immortal soul? Heroism in this sacrifice is measured precisely by madness: it is all the greater when it is offered for a more trivial motive.

Yet another temptation goes to the other extreme. With Sartre, it says: "*L'enfer, c'est les autres!*" ("Other people—that's hell!"). In that case, love itself becomes the great temptation and the great sin. Because it is an inescapable sin, it is also hell. But this too is only a disguised form of Eros—Eros in solitude. It is the love that is mortally wounded by its own incapacity to love another, and flies from others in order not to have to give itself to them. Even in its solitude this Eros is most tortured by its inescapable need of another, not for the other's sake but for its own fulfilment!

All these three answers are insufficient. The third says we must love only ourselves. The second says we must love only another. The first says that in loving another we simply seek the most effective way to love ourselves. The true answer, which is supernatural, tells us that we must love ourselves in order to be able to love others, that we must find ourselves by giving ourselves to them. The words of Christ are clear: "Thou shalt love thy neighbour as thyself."

This is not merely a helpful suggestion, it is the fundamental law of human existence. It forms part of the first and greatest commandment, and flows from the obligation to love God with all our heart and soul and strength. This double commandment, giving us two aspects of the same love, obliges us to another asceticism, which is not the answer of Eros, but the answer of Agapé.

Whatever may be said in following pages rests upon this foundation. Man is divided against himself and against God by his own selfishness, which divides him against his brother. This division cannot be healed by a love that places itself only on one side of the rift. Love must reach over to both sides and draw them together. We cannot love ourselves unless we love others, and we cannot love others unless we love ourselves. But a selfish love of ourselves makes us incapable of loving others. The difficulty of this commandment lies in the paradox that it would have us love ourselves unselfishly, because even our love of ourselves is something we owe to others.

This truth never becomes clear as long as we assume that each one of us, individually, is the centre of the universe. We do not exist for ourselves alone, and it is only when we are fully convinced of this fact that we begin to love ourselves properly and thus also love others. What do I mean by loving ourselves properly? I mean, first of all, desiring to live, accepting life as a very great gift and a great good, not because of what it gives us, but because of what it enables us to give to others. The modern world is beginning to discover, more and more, that the quality and vitality of a man's life depend on his own secret will to go on living. There is a dark force for destruction within us, which someone has called the "death instinct." It is a terribly powerful thing, this force generated by our own frustrated self-love battling with itself. It is the power of a self-love that has turned into self-hatred and which, in adoring itself, adores the monster by which it is consumed.

It is therefore of supreme importance that we consent to live not for ourselves but for others. When we do this we will be able first of all to face and accept our own limitations. As long as we secretly adore ourselves, our own deficiencies will remain to torture us with an apparent defilement. But if we live for others, we will gradually discover that no one expects us to be "as gods." We will see that we are human, like everyone else, that we all have weaknesses and deficiencies, and that these limitations of ours play a most important part in all our lives. It is because of them that we need others and others need us. We are not all weak in the same spots, and so we supplement and complete one another, each one making up in himself for the lack in another.

Only when we see ourselves in our true human context, as members of a race which is intended to be one organism and "one body," will we begin to understand the positive importance not only of the successes but of the failures and accidents in our lives. My successes are not my own. The way to them was prepared by others. The fruit of my labours is not my own: for I am preparing the way for the achievements of another. Nor are my failures my own. They may spring from the failure of another, but they are also compensated for by another's achievement. Therefore the meaning of my life is not to be looked for merely in the sum total of my own achievements. It is seen only in the complete integration of my achievements and failures with the achievements and failures of my own generation, and society, and time. It is seen, above all, in my integration in the mystery of Christ. That was what the poet John Donne realized during a serious illness when he heard the death knell tolling for another. "The Church is Catholic, universal," he said, "so are all her actions, all that she does belongs to all. . . . Who bends not his ear to any bell which upon any occasion rings? but who can remove it from that bell which is passing a piece of himself out of this world?"

Every other man is a piece of myself, for I am a part and a member of mankind. Every Christian is part of my own body, because we are members of Christ. What I do is also done for them and with them and by them. What they do is done in me and by me and for me. But each one of us remains responsible for his own share in the life of the whole body. Charity cannot be what it is supposed to be as long as I do not see that my life represents my own allotment in the life of a whole supernatural organism to which I belong. Only when this truth is absolutely central do other doctrines fit into their proper context. Solitude, humility, self-denial, action and contemplation, the sacraments, the monastic life, the family, war and peace—none of these make sense except in relation to the central reality which is God's love living and acting in those whom He has incorporated in His Christ. Nothing at all makes sense, unless we admit, with John Donne, that: "No man is an island, entire of itself; every man is a piece of the continent, a part of the main."

I

Love Can be Kept Only by Being Given Away

A HAPPINESS that is sought for ourselves alone can never be found: for a happiness that is diminished by being shared is not big enough to make us happy.

There is a false and momentary happiness in self-satisfaction, but it always leads to sorrow because it narrows and deadens our spirit. True happiness is found in unselfish love, a love which increases in proportion as it is shared. There is no end to the sharing of love, and, therefore, the potential happiness of such love is without limit. Infinite sharing is the law of God's inner life. He has made the sharing of ourselves the law of our own being, so that it is in loving others that we best love ourselves. In disinterested activity we best fulfil our own capacities to act and to be.

Yet there can never be happiness in compulsion. It is not enough for love to be shared: it must be shared freely. That is to say it must be given, not merely taken. Unselfish love that is poured out upon a selfish object does not bring perfect happiness: not because love requires a return or a reward for loving, but because it rests in the happiness of the beloved. And if the one loved receives love selfishly, the lover is not satisfied. He sees that his love has failed to make the beloved happy. It has not awakened his capacity for unselfish love.

Hence the paradox that unselfish love cannot rest perfectly except in a love that is perfectly reciprocated: because it knows that the only true peace is found in selfless love. Selfless love consents to be loved selflessly for the sake of the beloved. In so doing, it perfects itself.

The gift of love is the gift of the power and the capacity to love, and, therefore, to give love with full effect is also to receive it. So, love can only be kept by being given away, and it can only be given perfectly when it is also received.

2. Love not only prefers the good of another to my own, but it does not even compare the two. It has only one good: that of the beloved, which is, at the same time, my own. Love shares the good with another not by dividing it with him, but by identifying itself with him so that his good becomes my own. The same good is enjoyed in its wholeness by two in one spirit, not halved and shared by two souls. Where love is really disinterested, the lover does not even stop to inquire whether he can safely appropriate for himself some part of the good which he wills for his friend. Love seeks its whole good in the good of the beloved, and to divide that good would be to diminish love. Such a division would not only weaken the action of love, but in doing so would also diminish its joy. For love does not seek a joy that follows from its effect: its joy is in the effect itself, which is the good of the beloved. Consequently, if my love be pure I do not even have to seek for myself the satisfaction of loving. Love seeks one thing only: the good of the one loved. It leaves all the other secondary effects to take care of themselves. Love, therefore, is its own reward.

3. To love another is to will what is really good for him. Such love must be based on truth. A love that sees no distinction between good and evil, but loves blindly merely for the sake of loving, is hatred, rather than love. To love blindly is to love selfishly, because the goal of such love is not the real advantage of the beloved but only the exercise of love in our own souls. Such love cannot seem to be love unless it pretends to seek the good of the one loved. But since it actually cares nothing for the truth, and never considers that it may go astray, it proves itself to be selfish. It does not seek the true advantage of the beloved or even our own. It is not interested in the truth, but only in itself. It proclaims itself content with an apparent good: which is the exercise of love for its own sake, without any consideration of the good or bad effects of loving.

When such love exists on the level of bodily passion it is easily recognized for what it is. It is selfish, and, therefore, it is not love. Those whose love does not transcend the desires of their bodies, generally do not even bother to deceive themselves with good motives. They follow their passions. Since they do not deceive themselves, they are more honest, as well as more miserable, than those who pretend to love on a spiritual plane without realizing that their "unselfishness" is only a deception.

4. Charity is neither weak nor blind. It is essentially prudent, just, temperate, and strong. Unless all the other virtues blend together in charity, our love is not genuine. No one who really wants to love another will consent to love him falsely. If we are going to love others at all, we must make up our minds to love them well. Otherwise our love is a delusion.

The first step to unselfish love is the recognition that our love may be deluded. We must first of all purify our love by renouncing the pleasure of loving as an end in itself. As long as pleasure is our end, we will be dishonest with ourselves and with those we love. We will not seek their good, but our own pleasure.

5. It is clear, then, that to love others well we must first love the truth. And since love is a matter of practical and concrete human relations, the truth we must love when we love our brothers is not mere abstract speculation: it is the moral truth that is to be embodied and given life in our own destiny and theirs. This truth is more than the cold perception of an obligation, flowing from moral precepts. The truth we must love in loving our brothers is the concrete destiny and sanctity that are willed for them by the love of God. One who really loves another is not merely moved by the desire to see him contented and healthy and prosperous in this world. Love cannot be satisfied with anything so incomplete. If I am to love my brother, I must somehow enter deep into the mystery of God's love for him. I must be moved not only by human sympathy but by that divine sympathy which is revealed to us in Jesus and which enriches our own lives by the outpouring of the Holy Spirit in our hearts.

The truth I love in loving my brother cannot be something merely philosophical and abstract. It must be at the same time supernatural and concrete, practical and alive. And I mean these words in no metaphorical sense. The truth I must love in my brother is God Himself, living in him. I must seek the life of the Spirit of God breathing in him. And I can only discern and follow that mysterious life by the action of the same Holy Spirit living and acting in the depths of my own heart.

6. Charity makes me seek far more than the satisfaction of my own desires, even though they be aimed at another's good. It must also make me an instrument of God's Providence in their lives. I must become convinced and penetrated by the realization that without my love for them they may perhaps not achieve the things God has willed for them. My will must be the instrument of God's will in helping them create their destiny. My love must be to them the "sacrament" of the mysterious and infinitely selfless love God has for them. My love must be for them the minister not of my own spirit but of the Holy Spirit. The words I speak to them must be no other than the words of Christ who deigns to reveal Himself to them in me.

Such a conception of charity is, above all, proper to a priest. It is an aspect of the grace of Orders. It is, so to speak, inseparable from the priesthood, and a priest cannot be at peace with himself or with God unless he is trying to love others with a love that is not merely his but God's own love. Only this charity which is as strong and as sure as the Spirit of God Himself can save us from the lamentable error of pouring out on others a love that leads them into error and urges them to seek happiness where it can never be found.

7. In order to love others with perfect charity I must be true to them, to myself, and to God.

The true interests of a person are at once perfectly his own and common to the whole Kingdom of God. That is because these interests are all centred in God's designs for his soul. The destiny of each one of us is intended, by the Lord, to enter into the destiny of His entire Kingdom. And the more perfectly we are ourselves the more we are able to contribute to the good of the whole Church of God. For each person is perfected by the virtues of a child of God, and these virtues show themselves differently in everyone, since they come to light in the lives of each one of the saints under a different set of providential circumstances.

If we love one another truly, our love will be graced with a clear-sighted prudence which sees and respects the designs of God upon each separate soul. Our love for one another must be rooted in a deep devotion to Divine Providence, a devotion that abandons our own limited plans into the hands of God and seeks only to enter into the invisible work that builds His Kingdom. Only a love that senses the designs of Providence can unite itself perfectly to God's providential action upon souls. Faithful submission to God's secret working in the world will fill our love with piety, that is to say with supernatural awe and respect. This respect, this piety, gives our love the character of worship, without which our charity can never be quite complete. For love must not only *seek* the truth in the lives of those around us; it must *find* it there. But when we find the truth that shapes our lives we have found more than an idea. We have found a Person. We have come upon the actions of One who is still hidden, but whose work proclaims Him holy and worthy to be adored. And in Him we also find ourselves.

8. A selfish love seldom respects the rights of the beloved to be an autonomous person. Far from respecting the true

being of another and granting his personality room to grow and expand in its own original way, this love seeks to keep him in subjection to ourselves. It insists that he conform himself to us, and it works in every possible way to make him do so. A selfish love withers and dies unless it is sustained by the attention of the beloved. When we love thus, our friends exist only in order that we may love them. In loving them we seek to make pets of them, to keep them tame. Such love fears nothing more than the escape of the beloved. It requires his subjection because that is necessary for the nourishment of our own affections.

Selfish love often appears to be unselfish, because it is willing to make any concession to the beloved in order to keep him prisoner. But it is supreme selfishness to buy what is best in a person, his liberty, his integrity, his own autonomous dignity as a person, at the price of far lesser goods. Such selfishness is all the more abominable when it takes a complacent pleasure in its concessions, deluded that they are all acts of selfless charity.

A love, therefore, that is selfless, that honestly seeks the truth, does not make unlimited concessions to the beloved.

May God preserve me from the love of a friend who will never dare to rebuke me. May He preserve me from the friend who seeks to do nothing but change and correct me. But may He preserve me still more from one whose love is only satisfied by being rebuked.

If I love my brothers according to the truth, my love for them will be true not only to them but to myself.

I cannot be true to them if I am not true to myself.

"The Lord trieth the just and the wicked, but he that loveth iniquity hateth his own soul" (Psalm 10 : 6).

"Iniquity" is inequality, injustice, which seeks more for

myself than my rights allow and which gives others less than they should receive. To love myself more than others is to be untrue to myself as well as to them. The more I seek to take advantage of others the less of a person will I myself be, for the anxiety to possess what I should not have narrows and diminishes my own soul.

Therefore the man who loves himself too much is incapable of loving anyone effectively, including himself. How then can he hope to love another?

"An unjust man allureth his friend and leadeth him into a way that is not good" (Proverbs 16 : 29).

9. Charity must teach us that friendship is a holy thing, and that it is neither charitable nor holy to base our friendship on falsehood. We can be, in some sense, friends to all men because there is no man on earth with whom we do not have something in common. But it would be false to treat too many men as intimate friends. It is not possible to be intimate with more than very few, because there are only very few in the world with whom we have practically everything in common.

Love, then, must be true to the ones we love and to ourselves, and also to its own laws. I cannot be true to myself if I pretend to have more in common than I actually have with someone whom I may like for a selfish and unworthy reason.

There is, however, one universal basis for friendship with all men: we are all loved by God, and I should desire them all to love Him with all their power. But the fact remains that I cannot, on this earth, enter deeply into the mystery of their love for Him and of His love for them.

Great priests, saints like the Curé d'Ars, who have seen

into the hidden depths of thousands of souls, have, nevertheless, remained men with few intimate friends. No one is more lonely than a priest who has a vast ministry. He is isolated in a terrible desert by the secrets of his fellow men.

10. When all this has been said, the truth remains that our destiny is to love one another as Christ has loved us. Jesus had very few close friends when He was on earth, and yet He loved and loves all men and is, to every soul born into the world, that soul's most intimate friend. The lives of all the men we meet and know are woven into our own destiny, together with the lives of many we shall never know on earth. But certain ones, very few, are our close friends. Because we have more in common with them, we are able to love them with a special selfless perfection, since we have more to share. They are inseparable from our own destiny, and, therefore, our love for them is especially holy: it is a manifestation of God in our lives.

11. Perfect charity gives supreme praise to the liberty of God. It recognizes His power to give Himself to those who love Him purely without violating the purity of their love. More than that: selfless charity, by receiving from God the gift of Himself, becomes able, by that fact alone, to love with perfect purity. For God Himself creates the purity and the love of those who love Him and one another with perfect charity.

His charity must not be represented as hunger. It is the banquet of the Kingdom of Heaven, to which many were invited by the great King. Many could not come to the banquet because they desired something beyond it, something for themselves—a farm, a wife, a yoke of oxen.

They did not know that if they had sought first the banquet and the Kingdom they would have received everything else besides.

Charity is not hungry. It is the *juge convivium*—the perpetual banquet where there is no satiety, a feast in which we are nourished by serving others rather than by feeding ourselves. It is a banquet of prudence also, in which we know how to give to each other his just measure.

"And the Lord said: Who, thinkest thou, is the faithful and wise steward, whom his lord setteth over his family, to give them their measure of wheat in due season? Blessed is that servant, whom when his lord shall come, shall find him so doing" (Luke 12 : 43–44).

But to feed others with charity is to feed them with the Bread of Life, who is Christ, and to teach them also to love with a love that knows no hunger.

"I am the Bread of Life: he who comes to Me shall not hunger, and he who believes in Me shall never thirst" (John 6 : 35).

2

Sentences on Hope

We are not perfectly free until we live in pure hope. For when our hope is pure, it no longer trusts exclusively in human and visible means, nor rests in any visible end. He who hopes in God trusts God, whom he never sees, to bring him to the possession of things that are beyond imagination.

When we do not desire the things of this world for their own sake, we become able to see them as they are. We see at once their goodness and their purpose, and we become able to appreciate them as we never have before. As soon as we are free of them, they begin to please us. As soon as we cease to rely on them alone, they are able to serve us. Since we depend neither on the pleasure nor on the assistance we get from them, they offer us both pleasure and assistance, at the command of God. For Jesus has said: "Seek first the kingdom of God and His justice, and all these things [that is all that you need for your life on earth] will be given to you besides" (Matthew 6 : 33).

Supernatural hope is the virtue that strips us of all things in order to give us possession of all things. We do not hope for what we have. Therefore, to live in hope is to live in poverty, having nothing. And yet, if we abandon ourselves to economy of Divine Providence, we have everything we hope for. By faith we know God without seeing Him. By hope we possess God without feeling His presence.

If we hope in God, by hope we already possess Him, since hope is a confidence which He creates in our souls as secret evidence that He has taken possession of us. So the soul that hopes in God already belongs to Him, and to belong to Him is the same as to possess Him, since He gives Himself completely to those who give themselves to Him. The only thing faith and hope do not give us is the clear vision of Him whom we possess. We are united to Him in darkness, because we have to hope. *Spes quae videtur non est spes.**

Hope deprives us of everything that is not God, in order that all things may serve their true purpose as means to bring us to God.

Hope is proportionate to detachment. It brings our souls into the state of the most perfect detachment. In doing so, it restores all values by setting them in their right order. Hope empties our hands in order that we may work with them. It shows us that we have something to work for, and teaches us how to work for it.

Without hope, our faith gives us only an acquaintance with God. Without love and hope, faith only knows Him as a stranger. For hope casts us into the arms of His mercy and of His providence. If we hope in Him, we will not only come to know that He is merciful but we will experience His mercy in our own lives.

2. If, instead of trusting in God, I trust only in my own intelligence, my own strength, and my own prudence, the means that God has given to me to find my way to Him will all fail me. Nothing created is of any ultimate use

* "For we are saved by hope. But hope that is seen, is not hope. For what a man seeth, why doth he hope for?" (Romans 8 : 24).

without hope. To place your trust in visible things is to live in despair.

And yet, if I hope in God, I must also make a confident use of the natural aids which, with grace, enable me to come to Him. If He is good, and if my intelligence is His gift, then I must show my trust in His goodness by making use of my intelligence. I must let faith elevate, heal, and transform the light of my mind. If He is merciful, and if my freedom is a gift of His mercy, I must show my trust in His mercy by making use of my free will. I must let hope and charity purify and strengthen my human liberty and raise me to the glorious autonomy of a son of God.

Some who think they trust in God actually sin against hope because they do not use the will and the judgment He has given them. Of what use is it for me to hope in grace if I dare not make the act of will that corresponds with grace? How do I profit by abandoning myself passively to His will if I lack the strength of will to obey His commands? Therefore, if I trust in God's grace I must also show confidence in the natural powers He has given me, not because they are my powers but because they are His gift. If I believe in God's grace, I must also take account of my own free will, without which His grace would be poured out upon my soul to no purpose. If I believe that He can love me, I must also believe that I can love Him. If I do not believe I can love Him, then I do not believe Him who gave us the first commandment: "Thou shalt love the Lord thy God with thy whole heart and thy whole mind and all thy strength, and thy neighbour as thyself."

3. We can either love God because we hope for something from Him, or we can hope in Him knowing that He loves

us. Sometimes we begin with the first kind of hope and grow into the second. In that case, hope and charity work together as close partners, and both rest in God. Then every act of hope may open the door to contemplation, for such hope is its own fulfilment.

Better than hoping for anything from the Lord, besides His love, let us place all our hope in His love itself. This hope is as sure as God Himself. It can never be confounded. It is more than a promise of its own fulfilment. It is an effect of the very love it hopes for. It seeks charity because it has already found charity. It seeks God knowing that it has already been found by Him. It travels to Heaven realizing obscurely that it has already arrived.

4. All desires but one can fail. The only desire that is infallibly fulfilled is the desire to be loved by God. We cannot desire this efficaciously without at the same time desiring to love Him, and the desire to love Him is a desire that cannot fail. Merely by desiring to love Him, we are beginning to do that which we desire. Freedom is perfect when no other love can impede our desire to love God.

But if we love God for something less than Himself, we cherish a desire that can fail us. We run the risk of hating Him if we do not get what we hope for.

It is lawful to love all things and to seek them, once they become means to the love of God. There is nothing we cannot ask of Him if we desire it in order that He may be more loved by ourselves or by other men.

5. It would be a sin to place any limit upon our hope in God. We must love Him without measure. All sin is rooted in the failure of love. All sin is a withdrawal of love

from God, in order to love something else. Sin sets boundaries to our hope, and locks our love in prison. If we place our last end in something limited, we have withdrawn our hearts entirely from the service of the living God. If we continue to love Him as our end, but place our hope in something else together with Him, our love and our hope are not what they should be, for no man can serve two masters.

6. Hope is the living heart of asceticism. It teaches us to deny ourselves and leave the world not because either we or the world are evil, but because unless a supernatural hope raises us above the things of time we are in no condition to make a perfect use either of our own or of the world's true goodness. But we possess ourselves and all things in hope, for in hope we have them not as they are in themselves but as they are in Christ: full of promise. All things are at once good and imperfect. The goodness bears witness to the goodness of God. But the imperfection of all things reminds us to leave them in order to live in hope. They are themselves insufficient. We must go beyond them to Him in whom they have their true being.

We leave the good things of this world not because they are not good, but because they are only good for us in so far as they form part of a promise. They, in turn, depend on our hope and on our detachment for the fulfilment of their own destiny. If we misuse them, we ruin ourselves together with them. If we use them as children of God's promises, we bring them, together with ourselves, to God.

"For the expectation of the creature waiteth for the revelation of the sons of God. . . . Because the creature also itself shall be delivered from the servitude of corruption into the liberty of the glory of the children of God" (Romans 8 : 19–21).

Upon our hope, therefore, depends the liberty of the whole universe. Because our hope is the pledge of a new heaven and a new earth, in which all things will be what they were meant to be. They will rise, together with us, in Christ. The beasts and the trees will one day share with us a new creation and we will see them as God sees them and know that they are very good.

Meanwhile, if we embrace them for themselves, we discover both them and ourselves as evil. This is the fruit of the tree of the knowledge of good and evil—disgust with the things we have misused and hatred of ourselves for misusing them.

But the goodness of creation enters into the framework of holy hope. All created things proclaim God's fidelity to His promises, and urge us, for our sake and for their own, to deny ourselves and to live in hope and to look for the judgment and the general resurrection.

An asceticism that is not entirely suspended from this divine promise is something less than Christian.

7. The devil believes in God but he has no God. The Lord is not *his* God. To be at enmity with life is to have nothing to live for. To live forever without life is everlasting death: but it is a living and wakeful death without the consolation of forgetfulness. Now the very essence of this death is the absence of hope. The damned have confirmed themselves in the belief that they cannot hope in God. We sometimes think of the damned as men who think of only themselves as good, since all sin flows from pride

that refuses to love. But the pride of those who live as if they believed they were better than anyone else is rooted in a secret failure to believe in their own goodness. If I can see clear enough to realize that I am good because God has willed me to be good, I will at the same time be able to see more clearly the goodness of other men and of God. And I will be more aware of my own failings. I cannot be humble unless I first know that I am good, and know that what is good in me is not my own, and know how easy it is for me to substitute an evil of my own choice for the good that is God's gift to me.

8. Those who abandon everything in order to seek God know well that He is the God of the poor. It is the same thing to say that He is the God of the poor and that He is a jealous God—to say that He is a jealous God and a God of infinite mercy. There are not two gods, one jealous, whom we must fear, and one merciful, in whom we must place our hope. Our hope does not consist in pitting one of these gods against the other, bribing one to pacify the other. The Lord of all justice is jealous of His prerogative as the Father of mercy, and the supreme expression of His justice is to forgive those whom no one else would ever have forgiven.

That is why He is, above all, the God of those who can hope where there is no hope. The penitent thief who died with Christ was able to see God where the doctors of the law had just proved impossible Jesus's claim to divinity.

9. Only the man who has had to face despair is really convinced that he needs mercy. Those who do not want mercy never seek it. It is better to find God on the thres-

hold of despair than to risk our lives in a complacency that has never felt the need of forgiveness. A life that is without problems may literally be more hopeless than one that always verges on despair.

10. One of the greatest speculative problems in theology is resolved in practical Christian living by the virtue of hope. The mystery of free will and grace, of predestination and co-operation with God is resolved in hope which effectively co-ordinates the two in their right relation to one another. The one who hopes in God does not *know* that he is predestined to Heaven. But if he perseveres in his hope and continually makes the acts of will inspired by divine grace he will be among the predestined: for that is the object of his hope and "hope confoundeth not" (Romans 5 : 5). Each act of hope is his own free act, yet it is also a gift of God. And the very essence of hope is freely to expect all the graces necessary for salvation as free gifts from God. The free will that resolves to hope in His gifts recognizes, by that very fact, that its own act of hope is also His gift: and yet it also sees that if it did not will to hope, it would not let itself be moved by Him. Hope is the wedding of two freedoms, human and divine, in the acceptance of a love that is at once a promise and the beginning of fulfilment.

11. The faith that tells me God wills all men to be saved must be completed by the hope that God wills *me* to be saved, and by the love that responds to His desire and seals my hope with conviction. Thus hope offers the substance of all theology to the individual soul. By hope all the truths that are presented to the whole world in an abstract and

impersonal way become for me a matter of personal and intimate conviction. What I believe by faith, what I understand by the habit of theology, I possess and make my own by hope. Hope is the gateway to contemplation, because contemplation is an experience of divine things and we cannot experience what we do not in some way possess. By hope we lay hands on the substance of what we believe and by hope we possess the substance of the promise of God's love.

Jesus is the theology of the Father, revealed to us. Faith tells me that this theology is accessible to all men. Hope tells me that He loves me enough to give Himself to me. If I do not hope in His love for me, I will never really know Christ. I hear of Him by faith. But I do not achieve the contact that knows Him, and thereby knows the Father in Him, until my faith in Him is completed by hope and charity: hope that grasps His love for me and charity that pays Him the return of love I owe.

12. Hope seeks not only God in Himself, not only the means to reach Him, but it seeks, finally and beyond all else, God's glory revealed in ourselves. This will be the final manifestation of His infinite mercy, and this is what we pray for when we say "Thy Kingdom come."

3

Conscience, Freedom, and Prayer

To consider persons and events and situations only in the light of their effect upon myself is to live on the doorstep of hell. Selfishness is doomed to frustration, centred as it is upon a lie. To live exclusively for myself, I must make all things bend themselves to my will as if I were a god. But this is impossible. Is there any more cogent indication of my creaturehood than the insufficiency of my own will? For I cannot make the universe obey me. I cannot make other people conform to my own whims and fancies. I cannot make even my own body obey me. When I give it pleasure, it deceives my expectation and makes me suffer pain. When I give myself what I conceive to be freedom, I deceive myself and find that I am the prisoner of my own blindness and selfishness and insufficiency.

It is true, the freedom of my will is a great thing. But this freedom is not absolute self-sufficiency. If the essence of freedom were merely the act of choice, then the mere fact of making choices would perfect our freedom. But there are two difficulties here. First of all, our choices must really be free—that is to say they must perfect us in our own being. They must perfect us in our relation to other free beings. We must make the choices that enable us to fulfil the deepest capacities of our real selves. From this flows the second difficulty: we too easily assume that we *are* our real selves, and that our choices are really the ones we want to make when, in fact, our acts of free choice are (though morally imputable, no doubt) largely dictated by psychological compulsions, flowing from our inordinate ideas of our own importance. Our choices are too often dictated by our false selves.

Hence I do not find in myself the power to be happy merely by doing what I like. On the contrary, if I do nothing except what pleases my own fancy I will be miserable almost all the time. This would never be so if my will had not been created to use its own freedom in the love of others.

My free will consolidates and perfects its own autonomy by freely co-ordinating its action with the will of another. There is something in the very nature of my freedom that inclines me to love, to do good, to dedicate myself to others. I have an instinct that tells me that I am less free when I am living for myself alone. The reason for this is that I cannot be completely independent. Since I am not self-sufficient I depend on someone else for my fulfilment. My freedom is not fully free when left to itself. It becomes so when it is brought into the right relation with the freedom of another.

At the same time, my instinct to be independent is by no means evil. My freedom is not perfected by subjection to a tyrant. Subjection is not an end in itself. It is right that my nature should rebel against subjection. Why should my will have been created free, if I were never to use my freedom?

If my will is meant to perfect its freedom in serving another will, that does not mean it will find its perfection in serving *every* other will. In fact, there is only one will in whose service I can find perfection and freedom. To give my freedom blindly to a being equal to or inferior to myself is to degrade myself and throw away my freedom. I can only become perfectly free by serving the will of God. If I do, in fact, obey other men and serve them it is not for their sake alone that I will do so, but because their will is the sacrament of the will of God. Obedience to man has no meaning unless it is primarily obedience to God. From this flow many consequences. Where there is no faith in God there can be no real order; therefore, where there is no faith obedience is without any sense. It can only be imposed on others as a matter of expediency. If there is no God, no government is logical except tyranny. And in actual fact, states that reject the idea of God tend either to tyranny or to open disorder. In either case, the end is disorder, because tyranny is itself a disorder.

If I did not believe in God I think I would be bound in conscience to become an anarchist. Yet, if I did not believe in God, I wonder if I could have the consolation of being bound in conscience to do anything.

2. Conscience is the soul of freedom, its eyes, its energy, its life. Without conscience, freedom never knows what to do with itself. And a rational being who does not know what to do with himself finds the tedium of life unbearable. He is literally bored to death. Just as love does not find its fulfilment merely in loving blindly, so freedom wastes away when it merely "acts freely" without any purpose. An act without purpose lacks something of the perfection of freedom, because freedom is more than a matter of aimless choice. It is not enough to affirm my

liberty by choosing "something." I must use and develop my freedom by choosing something *good*.

I cannot make good choices unless I develop a mature and prudent conscience that gives me an accurate account of my motives, my intentions, and my moral acts. The word to be stressed here is *mature*. An infant, not having a conscience, is guided in its "decisions" by the attitude of somebody else. The immature conscience is one that bases its judgments partly, or even entirely, on the way other people seem to be disposed toward its decisions. The good is what is admired or accepted by the people it lives with. The evil is what irritates or upsets them. Even when the immature conscience is not entirely dominated by people outside itself, it nevertheless acts only as a representative of some other conscience. The immature conscience is not its own master. It is merely the delegate of the conscience of another person, or of a group, or of a party, or of a social class, or of a nation, or of a race. Therefore, it does not make real moral decisions of its own, it simply parrots the decisions of others. It does not make judgments of its own, it merely "conforms" to the party line. It does not really have motives or intentions of its own. Or if it does, it wrecks them by twisting and rationalizing them to fit the intentions of another. That is not moral freedom. It makes true love impossible. For if I am to love truly and freely, I must be able to give something that is truly my own to another. If my heart does not first belong to me, how can I give it to another? It is not mine to give!

3. Free will is not given to us merely as a firework to be shot off into the air. There are some men who seem to think their acts are freer in proportion as they are without

purpose, as if a rational purpose imposed some kind of limitation upon us. That is like saying that one is richer if he throws money out the window than if he spends it.

Since money is what it is, I do not deny that you may be worthy of all praise if you light your cigarettes with it. That would show you had a deep, pure sense of the ontological value of the pound. Nevertheless, if that is all you can think of doing with money you will not long enjoy the advantages that it can still obtain.

It may be true that a rich man can better afford to throw money out the window than a poor man: but neither the spending nor the waste of money is what makes a man rich. He is rich by virtue of what he has, and his riches are valuable to him for what he can do with them.

As for freedom, according to this analogy, it grows no greater by being wasted, or spent, but it is given to us as a talent to be traded with until the coming of Christ. In this trading we part with what is ours only to recover it with interest. We do not destroy it or throw it away. We dedicate it to some purpose, and this dedication makes us freer than we were before. Because we are freer, we are happier. We not only have more than we had but we become more than we were. This having and being come to us in a deepening of our union with the will of God. Our will is strengthened in obedience to the demands of objective reality. Our conscience is enlightened and it looks out upon a vastly widened horizon. We are able to see far nobler possibilities for the exercise of our freedom because we have grown in charity, and because we are enriched in divine grace we find in ourselves the power to attain ends that had been beyond us before.

All these fruits are meant to be gathered by our freedom

when we do the will of God. It is for this that we account ourselves happy when we know His will and do it, and realize that the greatest unhappiness is to have no sense of His purposes or His designs either for ourselves or for the rest of the world. "I walked at large," says the Psalmist, "because I have sought after thy commandments" (Psalm 118 : 45). "I have been delighted in the way of thy testimonies as in all riches. . . . Unless thy law had been my meditation, I had then perhaps perished in my abjection . . ." (Psalm 118 : 14, 92). "We are happy, O Israel, because the things that are pleasing to God are made known to us" (Baruch 4 : 4).

4. Our free acts must not only have a purpose, they must have the right purpose. And we must have a conscience that teaches us how to choose the right purposes. Conscience is the light by which we interpret the will of God in our own lives.

This light is twofold. First, there is the psychological conscience, which is better called consciousness. It reports to us the actions we perform. It is aware of them, and through them it is aware of itself. Second, there is our moral conscience, which tells us not only *that* we act, and *how* we act, but *how well* we act. It judges the value of our acts. The psychological and moral consciences are both faculties of the intelligence. They are two kinds of awareness of ourselves telling us what we really are.

Man is distinguished from the rest of creation by his intelligence and his freedom. He matures in his manhood by growing in wisdom and by gaining a more prudent and effective command of his own moral activity. Character and maturity are therefore measured by the clarity and discretion of our moral conscience. Conscience is the summary of the whole man, although a man is much more than an animated conscience. Conscience is the indication of hidden things, of imperceptible acts and tendencies that are much more important than itself. It is the mirror of a man's depths. The reality of a person is a deep and hidden thing, buried not only in the invisible recesses of man's own metaphysical secrecy but in the secrecy of God Himself.

Conscience is the face of the soul. Its changing expressions manifest more precisely the moral action of the soul than the changes of man's countenance manifest the emotions within him. Even the outward face of man is only a reflection of his conscience. True, only a very little of what is in a man's soul ever shines out in his face: but the little that is there is enough to speak eloquently of the conscience within.

5. One of the most important functions of the life of prayer is to deepen and strengthen and develop our moral conscience. The growth of our psychological conscience, although secondary, is not without importance also. The psychological conscience has its place in our prayer, but prayer is not the place for its proper development.

When we look inward and examine our psychological conscience our vision ends in ourselves. We become aware of our feelings, our inward activity, our thoughts, our judgments, and our desires. It is not healthy to be too constantly aware of all these things. Perpetual self-examination gives an over-anxious attention to movements that should remain instinctive and unobserved. When we attend too much to ourselves, our activity becomes cramped and stumbling. We get so much in our own way that we soon paralyse ourselves completely and become unable to act like normal human beings.

It is best, therefore, to let the psychological conscience alone when we are at prayer. The less we tinker with it the better. The reason why so many religious people believe they cannot meditate is that they think meditation consists in having religious emotions, thoughts, or affections of which one is oneself acutely aware. As soon as they start to meditate, they begin to look into the psychological conscience to find out if they are experiencing anything worthwhile. They find little or nothing. They either strain themselves to produce some interior experience, or else they give up in disgust.

6. The psychological conscience is most useful to us when it is allowed to act instinctively and without too much deliberate reflection on our own part. We should be able to see *through* our consciousness without seeing it at all. When the consciousness acts properly it is very valuable in prayer because it lends tone and quality to the action of the moral conscience, which is actually central in prayer.

At times the psychological conscience quickly gets paralysed under the stress of futile introspection. But there is another spiritual activity that develops and liberates its hidden powers of action: the perception of beauty. I do not mean by this that we must expect our consciousness to respond to beauty as an effete and esoteric thing. We ought to be alive enough to reality to see beauty all around us. Beauty is simply reality itself, perceived in a special way that gives it a resplendent value of its own. Everything, that is, is beautiful insofar as it is real—though the associations which they may have acquired for men may not always make things beautiful to us. Snakes are beautiful, but not to us.

One of the most important—and most neglected—elements in the beginnings of the interior life is the ability to respond to reality, to see the value and the beauty in ordinary things, to come alive to the splendour that is all around us in the creatures of God. We do not see these things because we have withdrawn from them. In a way we have to. In modern life our senses are so constantly bombarded with stimulation from every side that unless we developed a kind of protective insensibility we would go crazy trying to respond to *all* the advertisements at the same time!

The first step in the interior life nowadays is not, as some might imagine, learning *not* to see and taste and hear and feel things. On the contrary, what we must do is begin by unlearning our wrong ways of seeing, tasting, feeling, and so forth, and acquire a few of the right ones.

For asceticism is not merely a matter of renouncing television, cigarettes, and gin. Before we can begin to be ascetics, we first have to learn to see life as if it were something more than a hypnotizing telecast. And we must be able to taste something besides tobacco and alcohol: we must perhaps even be able to taste these luxuries themselves as if they too were good.

How can our conscience tell us whether or not we are renouncing things unless it first of all tells us that we know how to use them properly? For renunciation is not an end in itself: it helps us to use things better. It helps us to give them away. If reality revolts us, if we merely turn away from it in disgust, to whom shall we sacrifice it? How shall we consecrate it? How shall we make of it a gift to God and to men?

In an aesthetic experience, in the creation or the contemplation of a work of art, the psychological conscience is able to attain some of its highest and most perfect fulfilments. Art enables us to find ourselves and lose ourselves at the same time. The mind that responds to the intellectual and spiritual values that lie hidden in a poem, a painting, or a piece of music, discovers a spiritual vitality that lifts it above itself, takes it out of itself, and makes it present to itself on a level of being that it did not know it could ever achieve.

7. The soul that picks and pries at itself in the isolation of its own dull self-analysis arrives at a self-consciousness that is a torment and a disfigurement of our whole personality. But the spirit that finds itself above itself in the intensity and cleanness of its reaction to a work of art is "self-conscious" in a way that is productive as well as sublime. Such a one finds in himself totally new capacities for thought and vision and moral action. Without a moment of self-analysis he has discovered himself in discovering his capacity to respond to a value that lifts him above his normal level. His very response makes him better and different. He is conscious of a new life and new powers, and it is not strange that he should proceed to develop them.

It is important, in the life of prayer, to be able to respond to such flashes of aesthetic intuition. Art and prayer have never been conceived by the Church as enemies, and where the Church has been austere it has only been because she meant to insist on the essential difference between art and entertainment. The austerity, gravity, sobriety, and strength of Gregorian chant, of twelfth-century Cistercian architecture, of Carolingian minuscule script, have much to say about the life of prayer, and they have had much to do, in the past, with forming the prayer and the religious consciousness of saints. They have always done so in proportion as they have freed souls from concentration upon themselves, as well as from mere speculation about technical values in the arts and in asceticism. One can be at the same time a technical expert in chant and a man of prayer, but the moments of prayer and of technical criticism do not usually coincide.

If the Church has emphasized the function of art in her public prayer, it has been because she knew that a true and valid aesthetic formation was necessary for the wholeness of Christian living and worship. The liturgy and the chant and Church art are all supposed to form and spiritualize man's consciousness, to give him a tone and a maturity without which his prayer cannot normally be either very deep or very wide or very pure.

There is only one reason why this is completely true: art is not an end in itself. It introduces the soul into a higher spiritual order, which it expresses and in some sense explains. Music and art and poetry attune the soul to God because they induce a kind of contact with the Creator and Ruler of the Universe. The genius of the artist finds its way by the affinity of creative sympathy, or connaturality, into the living law that rules the universe. This law is nothing but the secret gravitation that draws all things to God as to their centre. Since all true art lays bare the action of this same law in the depths of our own nature, it makes us alive to the tremendous mystery of being, in which we ourselves, together with all other living and existing things, come forth from the depths of God and return again to Him. An art that does not produce something of this is not worthy of its name.

8. Before passing from the psychological conscience to the moral conscience, let us look at the subconscious mind. Too many religious people ignore the subconscious altogether. They either blithely suppose that it plays no part in their lives, or else they assume that it is simply an old attic that is not worth visiting, full of the rubbish from which we make our dreams.

It would be a great mistake to turn the interior life into a psychological experiment and make our prayer the object of psychoanalysis. If it is true and valid prayer, it needs no such analysis. But note that I have said *if*: for if it is not true prayer, it might very well benefit from analysis. The disheartening prevalence of false mysticism, the deadening grip that false asceticism sometimes gets on religious souls, and the common substitution of sentimentality for true religious feeling—all these things seem to warrant a little investigation of the subconscious substrate of what passes for "religion." However, that is by no means the province of the present book.

Only a few statements need be made here.

The subconscious mind plays a very important part in the interior life, even though it remains behind the scenes. Just as a good play depends on the scene, the lighting, and all the rest, so too our interior life owes much of its character to the setting and lighting and background and atmosphere which are provided, without any deliberate action of our own, by our subconscious mind.

In fact, it sometimes happens that the whole tone and atmosphere of a person's life of prayer—a certain emphasis on solitude or on sacrifice or on asceticism or on apostolic radiation—is provided by elements in the subconscious mind. For the subconscious mind is a store-house of images and symbols, I might almost say of "experiences" which provides us with more than half the material of what we actually experience as "life." Without our knowing it, we see reality through glasses coloured by the subconscious memory of previous experiences.

It is, therefore, important that our subconscious mind should enable us to live as our true selves. Indeed, it often happens that a man's true self is literally buried in the subconscious, and never has a chance to express itself except in symbolic protest against the tyranny of a malformed conscience that insists on remaining immature.

I do not say that we should try, without training or experience, to explore our own subconscious depths. But we ought at least to admit that they exist, and that they are important, and we ought to have the humility to admit we do not know all about ourselves, that we are not experts at running our own lives. We ought to stop taking our conscious plans and decisions with such infinite seriousness. It may well be that we are *not* the martyrs or the mystics or the apostles or the leaders or the lovers of God that we imagine ourselves to be. Our subconscious mind may be trying to tell us this in many ways—and we have trained ourselves, with the most egregious self-righteousness, to turn a deaf ear.

9. The psychological conscience is secondary in the life of the spirit. Although it can seize upon an occasional

reflection of ultimate reality, the psychological conscience cannot remain for long in union with a realm that is beyond our consciousness. But moral conscience can.

The moral conscience translates the general laws of being into the less general moral law, and, what is most important, it not only interprets the moral law to fit the circumstances of our own lives, but apprehends concretely, at every moment, that which is far more than any abstract norm of conduct. The moral conscience, by showing us the way of obedience to the inspirations of actual grace, *grasps and possesses at each moment of time the living law that is the will and love of God for ourselves.*

The distinction between the general abstract formulation of moral law, and the living, personal, concrete manifestation of God's will in our own lives is one of the most fundamental truths of Christianity, for it is the distinction between the letter, which kills, and the spirit, which gives life. Jesus, who came not to destroy the law but in order that every jot and tittle of the law should be fulfilled (Matthew 5 : 17–18), also taught that in order for the law to be fulfilled the doctors of the law would have to be confounded.

The justice of the scribes, who perfectly understood the letter of the law, was not sufficient to gain anyone admittance to the Kingdom of Heaven. It was necessary for the law to be fulfilled in spirit and in truth. It was necessary that men should be perfect in the law, not by the exterior observance of precepts but by the interior transformation of their whole being into sons of God. Then they would be children of their Father in Heaven, perfect as He is perfect (Matthew 5 : 45, 48). They would no longer keep the law with a formalistic perfection that defeated the whole purpose of the law, but they would realize that the sabbath was made for man, not man for the sabbath. They would cease to make void the law of God for the human traditions of ritualists and lawyers who could not understand Jesus when He taught that man must be born of the Holy Ghost in order to enter the Kingdom of God.*

In Christ we die to the letter of the law so that our conscience can no longer see things in the dead light of formalism and exterior observance. Our hearts refuse the dry husks of literal abstraction and hunger for the living bread and the eternal waters of the spirit which spring up to life everlasting.

"Therefore, my brethren, you are become dead to the Law by the body of Christ. . . . We are loosed from the law of death, wherein we were detained, so that we should serve in newness of spirit and not in the oldness of the letter" (Romans 7 : 4, 6). The law of life in the New Testament of Christ's grace is not merely a written document. It is the fulfilment, by charity, of God's designs in the consciences of those who answer the impulsions of His grace. The new law is not merely an exterior code of conduct but an interior life, the life of Jesus Himself, living by His spirit in those who remain united to Him by charity. The new law is expressed not only in the demands made upon us by divine and ecclesiastical precepts but above all by the exigencies of the Holy Spirit Himself, alive and active in the depths of our souls, constantly urging us to yield our wills to the gravitational pull of

* "Unless a man be born again of water and the Holy Ghost, he cannot enter the kingdom of God. That which is born of flesh is flesh, and that which is born of the Spirit is spirit. . . . Nicodemus answered and said: 'How can these things be done?' Jesus answered and said to him: 'Art thou a doctor in Israel and knowest not these things?' " (John 3 : 5, 6, 9–10).

charity, drawing us, through self-sacrifice, to the fulfilment of God's will in our own lives.

St. Paul knew that his own inspired writings were as nothing compared to the "writing" of Christ in the hearts of those who heard him. "You are the epistle of Christ," he told the Corinthians, "ministered by us and written not with ink, but with the Spirit of the living God. . . . [who] hath made us fit ministers of the New Testament not in the letter but in the spirit. For the letter killeth, but the spirit quickeneth" (II Corinthians 3 : 3, 6).

10. The whole function of the life of prayer is, then, to enlighten and strengthen our conscience so that it not only knows and perceives the outward, written precepts of the moral and divine laws, but above all lives God's law in concrete reality by perfect and continual union with His will. The conscience that is united to the Holy Spirit by faith, hope, and selfless charity becomes a mirror of God's own interior law which is His charity. It becomes perfectly free. It becomes its own law because it is completely subject to the will of God and to His Spirit. In the perfection of its obedience it "tastes and sees that the Lord is sweet," and knows the meaning of St. Paul's statement that the "law is not made for the just man" (I Timothy 1 : 9).

11. We do not have to create a conscience for ourselves. We are born with one, and no matter how much we may ignore it, we cannot silence its insistent demand that we do good and avoid evil. No matter how much we may deny our freedom and our moral responsibility, our intellectual soul cries out for a morality and a spiritual freedom without which it knows it cannot be happy. The first duty of

every man is to seek the enlightenment and discipline without which his conscience cannot solve the problems of life. And one of the first duties of society to the men who compose it is to enable them to receive the spiritual formation they need in order to live by the light of a prudent and mature conscience. I say "spiritual" and not merely "religious," for religious formation is sometimes no more than an outward formality, and therefore it is not really religious, nor is it a "formation" of the soul.

12. As a man is, so he prays. We make ourselves what we are by the way we address God. The man who never prays is one who has tried to run away from himself because he has run away from God. But unreal though he be, he is more real than the man who prays to God with a false and lying heart.

The sinner who is afraid to pray to God, who tries to deny God in his heart, is, perhaps, closer to confessing God than the sinner who stands before God, proud of his sin because he thinks it is a virtue. The former is more honest than he thinks, for he acknowledges the truth of his own state, confesses that he and God are not at peace with one another. The latter is not only a liar himself, but tries to make God a liar also, by calling upon Him to approve of his own lie. Such was the Pharisee in the parable, the holy man who practised many virtues, but who lied before God because he thought his piety made him better than other men. He despised sinners, and worshipped a false god who despised them like himself.

13. Prayer is inspired by God in the depth of our own nothingness. It is the movement of trust, of gratitude, of

adoration, or of sorrow that places us before God, seeing both Him and ourselves in the light of His infinite truth, and moves us to ask Him for the mercy, the spiritual strength, the material help that we all need. The man whose prayer is so pure that he never asks God for anything does not know who God is, and does not know who he is himself: for he does not know his own need of God.

All true prayer somehow confesses our absolute dependence on the Lord of life and death. It is, therefore, a deep and vital contact with Him whom we know not only as Lord but as Father. It is when we pray truly that we really *are*. Our being is brought to a high perfection by this, which is one of its most perfect activities. When we cease to pray, we tend to fall back into nothingness. True, we continue to exist. But since the main reason for our existence is the knowledge and love of God, when our conscious contact with Him is severed we sleep or we die. Of course, we cannot always, or even often, remain clearly conscious of Him. Spiritual wakefulness demands only the habitual awareness of Him which surrounds all our actions in a spiritual atmosphere without formally striking our attention except at certain moments of keener perception. But if God leaves us so completely that we are no longer disposed to think of Him with love, then we are spiritually dead.

Most of the world is either asleep or dead. The religious people are, for the most part, asleep. The irreligious are dead. Those who are asleep are divided into two classes, like the Virgins in the parable, waiting for the Bridegroom's coming. The wise have oil in their lamps. That is to say they are detached from themselves and from the cares of the world, and they are full of charity. They are indeed waiting for the Bridegroom, and they desire nothing else but His coming, even though they may fall asleep while waiting for Him to appear. But the others are not only asleep: they are full of other dreams and other desires. Their lamps are empty because they have burned themselves out in the wisdom of the flesh and in their own vanity. When He comes, it is too late for them to buy oil. They light their lamps only after He has gone. So they fall asleep again, with useless lamps, and when they wake up they trim them to investigate, once again, the matters of a dying world.

14. There are many levels of attention in prayer.

First of all, there is purely exterior attention. We "say prayers" with our lips, but our hearts are not following what we say although we think we would like to mean what we are saying. If we do not cultivate something better than this, we will seldom really pray. If we are quite content to pray without paying attention to our prayer or to God, it shows we have not much idea of who God is, and that we do not really appreciate the grace and the privilege of being able to speak to Him in prayer. For prayer is a gift of God, a gift which is by no means given to all men. Perhaps it is given to few because so few desire it, and of those who have received it so few have received it with gratitude.

At other times, we think of God in prayer but our thoughts of Him are not concerned with prayer. They are thoughts about Him that do not establish any contact with Him. So, while we pray, we are speculating about God and about the spiritual life, or composing sermons, or drawing up theological arguments. These thoughts are all right in their place, but if we take prayer seriously we will not call them prayer. For such thoughts cannot satisfy the soul that desires to find God in prayer. On the contrary, they leave it with a feeling of emptiness and dissatisfaction. At the same time, when one is really a man of prayer, speculative thoughts about God in the time of study or of intellectual work can often lead into prayer and give place to it; but only on condition that prayer is more to him than speculation.

Again, in prayer we are distracted by our practical difficulties, the problems of our state of life, the duties we have to face. It is not possible to avoid such distractions all the time, but if we know what prayer means, and know who God is, we will be able to turn these thoughts themselves into motives of prayer. But we will not be satisfied with such prayer as this. It is good, indeed, to turn distractions into material for petition, but it is better not to be distracted, or at least not to be drawn away from God by our distractions.

Then there is the prayer that is well used: words or thoughts serve their purpose and lead our minds and hearts to God, and in our prayer we receive light to apply these thoughts to our own problems and difficulties, to those of our friends, or to those of the Church. But sometimes this prayer, which is, of course, valid, leaves our hearts unsatisfied because it is more concerned with our problems, with our friends, and ourselves, than it is with God. However, if we are humble men, we will be grateful for ever so little light in our prayer, and will not complain too much, for it is a great thing to receive even a little light from so great a God.

There is a better way of prayer, a greater gift from God, in which we pass through our prayer to Him, and love Him. We taste the goodness of His infinite mercy. We know that we are indeed His sons, although we know our unworthiness to be called the sons of God. We know His infinite mercy in Jesus, and we know the meaning of the fact that we, who are sinners, indeed have a Saviour. And we learn what it is to know the Father in this Saviour, Jesus, His Son. We enter thus into a great mystery which cannot be explained, but only experienced. But in this prayer we still remain conscious of ourselves, we can reflect upon ourselves, and realize that we are the subjects of this great experience of love, as well as the objects of God's love.

In the beginning this reflexive quality in our prayer does not disturb us. But as we mature in the spiritual life it begins to be a source of unrest and dissatisfaction. We are ashamed to be so much aware of ourselves in our prayer. We wish we were not in the way. We wish our love for God were no longer spoiled and clouded by any return upon ourselves. We wish we were no longer aware that we rejoiced in His love, for we fear that our rejoicing might end in selfishness and self-complacency. And although we are grateful for the consolation and the light of His love, we wish we ourselves could disappear and see only Jesus. These two moments of prayer are like the two phases of the Apostles' vision of the Transfigured Christ on Mount Thabor. At first Peter, James, and John were delighted with the vision of Jesus, Moses, and Elias. They thought it would be a fine thing to build three tabernacles and stay there on the mountain for ever. But they were overshadowed by a cloud, and a voice came out of the cloud striking them with fear, and when they regained their vision they saw no one but Jesus alone.

So too there is another stage in our prayer, when consolation gives place to fear. It is a place of darkness and anguish and of conversion: for here a great change takes place in our spirit. All our love for God appears to us to have been full of imperfection, as indeed it has. We begin to doubt that we have ever loved Him. With shame and sorrow we find that our love was full of complacency, and that although we thought ourselves modest, we overflowed with conceit. We were too sure of ourselves, not afraid of illusion, not afraid to be recognized by other men as men of prayer. Now we see things in a different light, for we are in the cloud, and the voice of the Father fills our hearts with unrest and fear, telling us that we must no longer see ourselves: and yet, to our terror, Jesus does not appear to us and all that we see is—ourselves. Then what we find in our souls becomes terrible to us. Instead of complacently calling ourselves sinners (and secretly believing ourselves just), we begin to find that the sins of our past life were really sins, and really *our* sins—and we have not regretted them! And that since the time when we were grave sinners, we have still sinned without realizing it, because we were too sure we were the friends of God, and we have taken His graces lightly, or taken them to ourselves, and turned them to our own selfish profit, and used them for our own vanity, and even exploited them to lift ourselves above other men, so that in many ways we have turned the love of God into selfishness and have revelled in His gifts without thanking Him or using them for His glory.

Then we begin to see that it is just and right that we be

abandoned by God, and left to face many and great temptations. Nor do we complain of these temptations, for we are forced to recognize that they are only the expression of the forces that were always hiding behind the facade of our supposed virtues. Dark things come out of the depths of our souls, and we have to consider them and recognize them for our own, and then repudiate them, lest we be saddled with them for eternity. Yet they return, and we cannot escape them. They plague us in our prayer. And while we face them, and cannot get rid of them, we realize more clearly than ever before our great need for God, and the tremendous debt we owe His honour, and we try to pray to Him and it seems that we cannot pray. Then begins a spiritual revaluation of all that is in us. We begin to ask ourselves what is and is not *real* in our ideals!

This is the time when we really learn to pray in earnest. For now we are no longer proud enough to expect great lights and consolations in our prayer. We are satisfied with the driest crust of supernatural food, glad to get anything at all, surprised that God should even pay the slightest attention. And if we cannot pray (which is a source of concern) yet we know more than ever before how much we desire to pray. If we could be consoled at all, this would be our only consolation.

The man who can face such dryness and abandonment for a long time, with great patience, and ask nothing more of God but to do His holy will and never offend Him, finally enters into pure prayer. Here the soul goes to God in prayer without any longer adverting either to itself or to its prayer. It speaks to Him without knowing what it is saying because God Himself has distracted the mind from its words and thoughts. It reaches Him without thoughts because, before it can think of Him, He is already present in the depths of the spirit, moving it to love Him in a way it cannot explain or understand. Time no longer means anything in such prayer, which is carried on in instants of its own, instants that can last a second or an hour without our being able to distinguish one from another. For this prayer belongs less to time than to eternity.

This deep interior prayer comes to us of its own accord, that is, by the secret movement of the Spirit of God, at all times and in all places, whether we be praying or not. It can come at work, in the middle of our daily business, at a meal, on a silent road, or in a busy thoroughfare, as well as at Mass, or in church, or when we recite the psalms in choir. However, such prayer draws us naturally to interior and even exterior solitude. It does not depend on exterior conditions, but it has effected such an interior isolation and solitariness in our own souls that we naturally tend to seek silence and solitude for our bodies as well as for our souls. And it is good for the soul to be in solitude for a great part of the time. But if it should seek solitude for its own comfort and consolation, it will have to endure more darkness and more anguish and more trial. Pure prayer only takes possession of our hearts for good when we no longer desire any special light or grace or consolation for ourselves, and pray without any thought of our own satisfaction.

Finally, the purest prayer is something on which it is impossible to reflect until after it is over. And when the grace has gone we no longer seek to reflect on it, because we realize that it belongs to another order of things, and that it will be in some sense debased by our reflecting on it. Such prayer desires no witness, even the witness of our own souls. It seeks to keep itself entirely hidden in God.

The experience remains in our spirit like a wound, like a scar that will not heal. But we do not reflect upon it. This living wound may become a source of knowledge, if we are to instruct others in the ways of prayer; or else it may become a bar and an obstacle to knowledge, a seal of silence set upon the soul, closing the way to words and thoughts, so that we can say nothing of it to other men. For the way is left open to God alone. This is like the door spoken of by Ezechiel, which shall remain closed because the King is enthroned within.

4

Pure Intention

If God were merely another contingent being like myself, then to do His will would seem to be just as futile as doing my own. Our happiness consists in doing the will of God. But the essence of this happiness does not lie merely in an agreement of wills. It consists in a union with God. And the union of wills which makes us happy in God must ultimately be something deeper than an agreement.

2. First of all, let us not all be too glib in our statements about the will of God. God's will is a profound and holy mystery, and the fact that we live our everyday lives engulfed in this mystery should not lead us to underestimate its holiness. We dwell in the will of God as in a sanctuary. His will is the cloud of darkness that surrounds His immediate presence. It is the mystery in which His divine life and our created life become "one spirit," since, as St. Paul says, "Those who are joined to the Lord are one spirit" (I Corinthians 6 : 17).

There are religious men who have become so familiar with the concept of God's will that their familiarity has bred an apparent contempt. It has made them forget that God's will is more than a concept. It is a terrible and transcendent reality, a secret power which is given to us, from moment to moment, to be the life of our life and the soul of our own soul's life. It is the living flame of God's own

Spirit, in whom our own soul's flame can play, if it wills, like a mysterious angel. God's will is not an abstraction, not a machine, not an esoteric system. It is a living concrete reality in the lives of men, and our souls are created to burn as flames within His flame. The will of the Lord is not a static centre drawing our souls blindly toward itself. It is a creative power, working everywhere, giving life and being and direction to all things, and above all forming and creating, in the midst of an old creation, a whole new world which is called the Kingdom of God. What we call the "will of God" is the movement of His love and wisdom, ordering and governing all free and necessary agents, moving movers and causing causes, driving drivers and ruling those who rule, so that even those who resist Him carry out His will without realizing that they are doing so. In all His acts God orders all things, whether good or evil, for the good of those who know Him and seek Him and who strive to bring their own freedom under obedience to His divine purpose. All that is done by the will of God in secret is done for His glory and for the good of those whom He has chosen to share in His glory.

3. Shall I be content to do God's will for my own advantage?

It is better to do His will with a weak but deliberate co-operation than to do His will unconsciously, unwillingly, and in spite of myself. But let me not confine my idea of perfection to the selfish obedience that does God's will merely for the sake of my own profit. True happiness is not found in any other reward than that of being united with God. If I seek some other reward besides God Himself, I may get my reward but I cannot be happy.

The secret of pure intention is not to be sought in the renunciation of all advantage for ourselves. Our intentions are pure when we identify our advantage with God's glory, and see that our happiness consists in doing His will because His will is right and good. In order to make our intentions pure, we do not give up all idea of seeking our own good, we simply seek it where it can really be found: in a good that is beyond and above ourselves. Pure intention identifies our own happiness with the common good of all those who are loved by God. It seeks its joy in God's own will to do good to all men in order that He may be glorified in them.

And, therefore, a pure intention is actually the most efficacious way of seeking our own advantage and our own happiness.

4. An impure intention is one that yields to the will of God while retaining a preference for my own will. It divides my will from His will. It gives me a choice between two advantages: one in doing His will and one in doing my own. An impure intention is imprudent, because it weighs truth in the balance against illusion; it chooses between a real and an apparent good as if they were equal.

A pure intention sees that the will of God is always good. An impure intention, without doubting in theory that God wills what is universally best, practically doubts that He can always will what is best for me in willing what is best for all. And so the man whose intention is not pure is compelled by his own weakness and imprudence to pass judgment on the will of God before he obeys it. He is not free to do the will of God with perfect generosity. He diminishes his love and his obedience by making an adjustment between God's will and his own, and so the will of God comes to have, for him, a variety of values: richer when it is more pleasing to him, poorer when it offers less immediate satisfaction, valueless when it demands a sacrifice of his own selfish interests.

5. Only a pure intention can be clear-sighted and prudent. The man of impure intentions is hesitant and blind. Since he is always caught between two conflicting wills, he cannot make simple and clear-cut decisions. He has twice as much to think about as the man who seeks only the will of God, since he has to worry about his own will and God's will at the same time. He cannot be really happy, because happiness is impossible without interior freedom, and we do not have interior freedom to do what we please without anxiety, unless we take pleasure in nothing but the will of God.

6. The man of impure intentions may not clearly realize that he is deceiving himself. Blinded by his own selfishness, he cannot even see that he is blinded. The hesitation that divides him between God's will and his own is by no means clear. It does not involve a practical choice between two clearly seen alternatives. It plunges him into a confusion of doubtful choices, a welter of possibilities. If he had enough interior peace to listen to his own conscience, he would hear it telling him that he does not really know what he is doing. He realizes obscurely that if he knew himself better he would be less likely to deceive himself. He knows that he is blindly following his own selfish ideas, under cover of motives he has not taken time to examine. But he does not really want to examine them,

because if he did so he might find out that his will and the will of God were directly opposed to one another. He might discover that there was no alternative for him but to do the will of God, which he does not really want to do.

7. Sanctity does not consist merely in *doing* the will of God. It consists in *willing* the will of God. For sanctity is union with God, and not all those who carry out His will are united with His will. Even those who commit sin contribute, by the effects of their sin, to the fulfilment of the will of God. But because they sin they formally will what God does not will. And a man can also sin by failing to will what God wills him to do. In either case, he may do what God wills while himself willing the opposite.

It is not always necessary to find out what God wills in order to do it. A man can live like a tree or an animal, doing the divine will all his life and never knowing anything about it. But if we are to will what He wills we must begin to know something of what He wills. We must at least desire to know what He wills.

If the Lord has given me intelligence, it is because He wills me to see something of His intentions for me, in order that I may enter into His plans with a free and intelligent co-operation. And so I cannot merely shut my eyes and will "whatever He wills" without ever looking up to see what He is doing.

It is true that we do not always know what the will of God for us really is. Perhaps we know it far less often than we imagine. That does not mean that we must not seek to know it. He wills that we obey in everything that we know to be commanded by Him, that we do nothing that He has forbidden, that we will all that He wills us to will and reject all that He wills us to reject. After that we must solve all our doubts by testing them with His known will, and by doing what is uncertain only in the light of what is certainly His will.

8. How can I find out what is the will of God for me?

Before the Lord wills me to do anything, He first of all wills me to *be*. What I do must depend on what I am. Therefore, my being itself contains in its own specific nature a whole code of laws, ways of behaving, that are willed for me by the God who has willed me to be.

My rational nature as a free and intelligent being postulates that I guide my actions not by blind instinct but by reason and free choice. But if God has willed me to be a man, and if the response of my manhood to His divine command is an act of will, my fundamental homage to the creative will of God is the will, on my own part, to be the man He wants me to be.

A man is only perfectly a man when he consents to live as a son of God. The consent to live as a son implies the consciousness of a divine inheritance: "If we are sons, heirs also and joint heirs with Christ" (Romans 8 : 17).

It is the will of God that we live not only as rational beings, but as "new men" regenerated by the Holy Spirit in Christ. It is His will that we reach out for our inheritance, that we answer His call to be His sons. We are born men without our own consent, but the consent to be sons of God has to be elicited by our own free will. We are obliged to learn what this consent consists of, and we find that it is an act of faith in Christ, by which we receive into our hearts the Spirit of God. The Holy Spirit is the One who makes us sons of God, justifying our souls by His presence and His charity, granting us the power to live and act as sons of God. "For the Spirit Himself giveth testimony to our spirit that we are the sons of God" (Romans 8 : 16).

Now the divine inheritance which God the Father gives to us in the Spirit of His love is simply the life of His incarnate Word in our souls. If we would live like sons of God, we must reproduce in our own lives the life and the charity of His only begotten Son. We must, therefore, live by the commandments and the counsels and by the Spirit of Jesus. And in order to do this we must search the Scriptures and understand the Gospels, in order to find out what Jesus is like and what His commandments are.

Besides that, we have to seek Him where He is to be found living among us on earth: in the Kingdom He came to establish, which is His Church. We must listen to His voice not only in the Scriptures but in the authority which, as we read in the Scriptures, He constituted over us to rule and sanctify and teach us by His own light, and His own holiness, and His own power. It was to the Apostles and to their successors that Jesus said: "The Paraclete, the Holy Ghost whom the Father will send in my name, He will teach you all things and bring all things to your mind whatsoever I shall have said to you" (John 14 : 26).

We receive the Holy Spirit through the Church and her sacraments. The Church, herself guided and enlivened and formed by the Spirit of Jesus, forms Christ in our souls, and gives us His life in giving us His Holy Spirit, that we may know Him as she knows Him and that we may be united to Him as she is, in the bonds of perfect charity and in the wisdom of contemplation.

If we have His Spirit in our hearts, we will be urged by the charity of Christ to live in charity and self-sacrifice like Jesus, who said: "Whosoever doth not carry his Cross and come after me cannot be my disciple" (Luke 14 : 26).

9. The Spirit of God makes Himself known in our hearts by awakening in us the recognition of God's love for us in His Son Jesus Christ, and by showing us how to keep His commandments. "By this hath the charity of God appeared towards us, because God hath sent His only begotten Son into the world, that we may live by Him. . . . By this is the Spirit of God known: Every spirit which confesseth that Jesus Christ is come in the flesh, is of God. . . . By this we know the spirit of truth and the spirit of error. . . . Everyone that loveth is born of God. He that loveth not, knoweth not God, for God is charity" (I John 4).

Above all, the Holy Spirit teaches us to live, not according to the flesh but according to divine charity. "If you live by the flesh you shall die, but if by the spirit you mortify the deeds of the flesh, you shall live" (Romans 8 : 13). "Now the works of the flesh are manifest which are fornication, uncleanness, immodesty, luxury, idolatry, witchcrafts, enmities, contentions, emulations, wraths, quarrels, dissensions, sects, envies, murders, drunkenness, revellings and such like. Of which I foretell you as I have foretold to you that they who do such things shall not obtain the Kingdom of God" (Galatians 5 : 19–21).

If we have the Spirit of God in our hearts, we will live by His law of charity, inclined always to peace rather than dissension, to humility rather than arrogance, to obedience rather than rebellion, to purity and temperance, to simplicity and quietness and calm, to strength, generosity, and wisdom, to prudence and all-embracing justice, and we will love others more than ourselves, for it is the commandment of Jesus that we should love one another as He has loved us (John 15 : 12).

None of these things can be done without prayer, and we must turn to prayer first of all, not only to discover God's will but above all to gain the grace to carry it out with all the strength of our desire.

10. The will of God, which the Spirit of God Himself teaches us in the secrecy of our inmost being, must always remain as much of a mystery as God Himself. Our desire to know His will implies rather a desire to recognize certain signs of the mystery of His will, than to penetrate the mystery in itself. If we do not remember this distinction we no longer revere the holiness and the mystery of God's will in itself. We judge the invisible reality of His will by the visible and sometimes contemptible signs which show us where His will is found.

When we speak of God's will, we are usually speaking only of some recognizable sign of His will. The signpost that points to a distant city is not the city itself, and sometimes the signs that point to a great place are in themselves insignificant and contemptible. But we must follow the direction of the signpost if we are to get to the end of our journey.

Everything that exists and everything that happens bears witness to the will of God. It is one thing to see a sign and another thing to interpret that sign correctly. However, our first duty is to recognize signs for what they are. If we do not even regard them as indications of any-

thing beyond themselves, we will not try to interpret them.

Of all the things and all the happenings that proclaim God's will to the world, only very few are capable of being interpreted by men. And of these few, fewer still find a capable interpreter. So that the mystery of God's will is made doubly mysterious by the signs that veil it from our eyes. To know anything at all of God's will we have to participate, in some manner, in the vision of the prophets: men who were always alive to the divine light concealed in the opacity of things and events, and who sometimes saw glimpses of that light where other men saw nothing but ordinary happenings.

And yet if we are too anxious to pry into the mystery that surrounds us we will lose the prophet's reverence and exchange it for the impertinence of soothsayers. We must be silent in the presence of signs whose meaning is closed to us. Otherwise we will begin incontinently to place our own superstitious interpretation upon everything—the number of steps to a doorway, a card pulled out of the pack, the shadow of a ladder, the flight of birds. God's will is not so cheap a mystery that it can be unlocked by any key like these!

Nevertheless, there are some signs that everyone must know. They must be easily read and seen, and they are indeed very simple. But they come sparingly, few in number; they show us clearly enough the road ahead but not for more than a few paces. When we have taken those few paces, what will happen? We must learn to be poor in our dependence on these clear signs, to take them as they come, not to demand more of them than we need, not to make more of them than they really tell.

If I am to know the will ot God, I must have the right attitude toward life. I must first of all know what life is, and to know the purpose of my existence.

It is all very well to declare that I exist in order to save my soul and give glory to God by doing so. And it is all very well to say that in order to do this I obey certain commandments and keep certain counsels. Yet knowing this much, and indeed knowing all moral theology and ethics and canon law, I might still go through life conforming myself to certain indications of God's will without ever fully giving myself to God. For that, in the last analysis, is the real meaning of His will. He does not need our sacrifices, He asks for our *selves.* And if He prescribes certain acts of obedience, it is not because obedience is the beginning and the end of everything. It is only the beginning. Charity, divine union; transformation in Christ: these are the end.

So let me clearly realize first of all that what God wants of me is myself. That means to say that His will for me points to one thing: the realization, the discovery, and the fulfilment of my self, my true self, in Christ. And that is why the will of God so often manifests itself in demands that I sacrifice myself. Why? Because in order to find my true self in Christ, I must go beyond the limits of my own narrow egoism. In order to save my life, I must lose it. For my life in God is and can only be a life of unselfish charity.

When Jesus said "He that would save his life will lose it, and he that would lose his life for my sake shall find it," He was teaching us the great truth that God's will for us is, before all else, that we should find ourselves, find our true life, or, as the Vulgate text has it, find our *souls*. God's will for us is not only that we should be the persons He means us to be, but that we should share in His work of creation and *help Him to make us* into the persons He means us to be. Always, and in all things, God's will for me is that I should shape my own destiny, work out my own salvation, forge my own eternal happiness, in the way He has planned it for me. And since no man is an island, since we all depend on one another, I cannot work out God's will in my own life unless I also consciously help other men to work out His will in theirs. His will, then, is our sanctification, our transformation in Christ, our deeper and fuller integration with other men. And this integration results not in the absorption and disappearance of our own personality, but in its affirmation and its perfection.

Everything that God wills in my life is directed to this double end: my perfection as part of a universal whole, and my perfection in myself as an individual person, made in God's image and likeness. The most important part of man's education is the formation of a conscience that is capable of seeing God's will in this correct light, and guiding the response of his own will in strong, prudent, and loving decisions. So to live is true wisdom.

This view of life as a growth in God, as a transformation in Christ, and as a supernatural self-realization in the mystical body of Christ is the only one that really helps us to recognize and interpret the will of God correctly. Without this view of life, we will not even be able to see the most obvious manifestations of the divine will—the manifestations that are made clear to us in the ordinary circumstances of our everyday life. For in the course of each day the duties of our state, the claims made on us by those around us, the demands on our energy, our patience, and our time, all make known to us the will of God and show us the way to realize ourselves in Him by losing ourselves in charity. But the pharisee who splits hairs and rationalizes his way out of these chances for self-dedication, although he may theorize and dogmatize about the will of God, never fully does that will for he never really abandons himself to the influence of divine charity.

Of such men God spoke through the prophet Isaias, saying: "For they seek me from day to day, and desire to know my ways as a nation that hath done justice and hath not forsaken the judgment of their God; they ask of me judgments of justice: they are willing to approach to God. Why have we fasted, and thou hast not regarded: have we humbled our souls and thou hast not taken notice? Behold in the day of your fast your own will is found, and you exact of all your debtors" (Isaias 58 : 2–3).

11. Wherever we have some sign of God's will, we are obliged to conform to what the sign tells us. We should do so with a pure intention, obeying God's will because it is good in itself as well as good for us. It takes more than an occasional act of faith to have such pure intentions. It takes a whole life of faith, a total consecration to hidden values. It takes sustained moral courage and heroic confidence in the help of divine grace. But above all it takes humility and spiritual poverty to travel in darkness and uncertainty, where so often we have no light and see no sign at all.

12. If sanctity consists in willing the will of God, as well as in doing it, perfect sanctity consists in willing the will of God perfectly. But absolute perfection is not possible,

in this matter, to any man on earth. In order to will perfectly what God wills, we would have to know as perfectly as He does what He wills. Our perfection will consist in explicitly willing whatever of God's will is certain to us, implicitly willing all that we do not know, and doing all this with the motive that is best and most perfect for us.

It is not enough to do the will of God because His will is unavoidable. Nor is it enough to will what He wills because we have to. We have to will His will because we love it. Yet it would be a false idea of perfection for an imperfect person suddenly to try to act with a perfection he does not possess. It is not the will of God that we should obey Him while at the same time telling Him lies about our interior dispositions. If our dispositions are bad, let us ask Him to make them better, but let us not tell Him that they are really very good. Still less is it enough to say, "Thy will be done," and then do the opposite. It is better to be like the son in the parable, who said, "I will not" (Matthew 21 : 28), but afterward went to work in the vineyard, than to be like the other one who said, "I go, sir," and then did not obey.

13. If, in trying to do the will of God, we always seek the highest abstract standard of perfection, we show that there is still much we need to learn about the will of God. For God does not demand that every man attain to what is theoretically highest and best. It is better to be a good street sweeper than a bad writer, better to be a good bartender than a bad doctor, and the repentant thief who died with Jesus on Calvary was far more perfect than the holy ones who had Him nailed to the cross. And yet, abstractly

speaking, what is more holy than the priesthood and less holy than the state of a criminal? The dying thief had, perhaps, disobeyed the will of God in many things: but in the most important event of his life he listened and obeyed. The Pharisees had kept the law to the letter and had spent their lives in the pursuit of a most scrupulous perfection. But they were so intent upon perfection as an abstraction that when God manifested His will and His perfection in a concrete and definite way they had no choice but to reject it.

14. Let me then wish to do God's will because it is His will. Let me not seek to measure His will by some abstract standard of perfection outside Himself. His will is measured by the infinite reality of His love and wisdom, with which it is identified. I do not have to ask if His will be wise, once I know it is His.

If I do His will as a free act of homage and adoration paid to a wisdom that I cannot see, His will itself becomes the life and substance and reality of my worship. But if I do His will as a perfunctory adjustment of my own will to the unavoidable, my worship is hollow and without heart.

15. The perfect love of God's will is a union so close that God Himself both utters and fulfils His will at the same instant in the depths of my own soul. Pure intention, in this highest sense, is a secret and spiritual word of God which not only commands my will to act, or solicits my co-operation, but fulfils what He says in me. The action is at once perfectly mine and perfectly His. But its substance comes entirely from Him. In me, it is entirely received:

only to be offered back to Him in the silent ovation of His own inexpressible love. Such words, such "intentions," which at once seek and find what they seek and give it back to God, resound with power and authority through all my faculties, so that my entire being is transformed into an expression and a fulfilment of what they say. This is what Jesus meant when He said that it was His "food" to do the will of the Father who sent Him. The will of God, accomplished as it is uttered, identifying us at once with Him who speaks and with what He says in us, makes our entire being a perfect reflection of Him who desires to see His will done in our hearts. The thing willed is not important, only He who wills it, for He cannot will anything that is against His will, that is to say, against His wisdom and His perfection.

Once we have heard the voice of the Almighty fulfilling His own command by speaking it in our hearts, we realize that our contemplation can never again be a mere looking or a mere seeking: it must also be a doing and a fulfilment. We hunger for the transforming words of God, words spoken to our spirit in secret and containing our whole destiny in themselves. We come to live by nothing but this voice. Our contemplation is rooted in the mystery of Divine Providence, and in its actuality. Providence can no longer be for us a philosophical abstraction. It is no longer a supernatural agency to provide us with food and clothing at the right time. Providence itself becomes our food and our clothing. God's mysterious decisions are themselves our life.

16. This fact tends to resolve the antinomy between action and contemplation. "Action" is no longer a matter of

resigning ourselves to works that seem alien to our life in God: for the Lord Himself places us exactly where He wants us to be and He Himself works in us. "Contemplation" is no longer merely the brief, satisfying interlude or reward in which our works are relieved by recollection and peace. Action and contemplation now grow together into one life and one unity. They become two aspects of the same thing. Action is charity looking outward to other men, and contemplation is charity drawn inward to its own divine source. Action is the stream, and contemplation is the spring. The spring remains more important than the stream, for the only thing that really matters is for love to spring up inexhaustibly from the infinite abyss of Christ and of God.

It is for us to take care that these living waters well up in our own hearts. God will make it His own concern to guide our action, if we live in Him, and He will turn the stream into whatever channels He wills. "As the divisions of the waters, so the heart of the king is in the hand of the Lord: whithersoever He will, He shall turn it" (Proverbs 21 : 1).

17. When action and contemplation dwell together, filling our whole life because we are moved in all things by the Spirit of God, then we are spiritually mature. Our intentions are habitually pure. Johannes Tauler somewhere makes a distinction between two degrees of pure intention, one of which he calls *right* intention, and the other *simple* intention. They may serve to explain the union of action and contemplation in one harmonious whole.

When we have a *right* intention, our intention is pure. We seek to do God's will with a supernatural motive.

We mean to please Him. But in doing so we still consider the work and ourselves apart from God and outside Him. Our intention is directed chiefly upon the work to be done. When the work is done, we rest in its accomplishment, and hope for a reward from God.

But when we have a *simple* intention, we are less occupied with the thing to be done. We do all that we do not only for God but so to speak *in* Him. We are more aware of Him who works in us than of ourselves or of our work. Yet this does not mean that we are not fully conscious of what we do, or that realities lose their distinctness in a kind of sweet metaphysical blur. It may happen that one who works with this "simple" intention is more perfectly alive to the exigencies of his work and does the work far better than the worker of "right" intention who has no such perspective. The man of right intentions makes a juridical offering of his work to God and then plunges himself into the work, hoping for the best. For all his right intention he may well become completely dizzy in a maze of practical details.

A right intention demands that we work with enough detachment to keep ourselves *above* the work to be done. But it does not altogether prevent us from gradually sinking into it over our ears. When this happens, we have to pull ourselves out, leave the work aside, and try to recover our balance and our right intention in an interval of prayer.

The man of simple intention, because he is essentially a contemplative, works always in an atmosphere of prayer. I do not say merely that he works in an atmosphere of peace. Anyone who works sanely at a job he likes can do as much as that. But the man of simple intention works in an atmosphere of prayer: that is to say he is recollected. His spiritual reserves are not all poured out into his work, but stored where they belong, in the depths of his being, with his God. He is detached from his work and from its results. Only a man who works purely for God can at the same time do a very good job and leave the results of the job to God alone. If our intention is less than simple, we may do a very good job, but in doing so we will become involved in the hope of results that will satisfy ourselves. If our intention is less than right we will be concerned neither for the job nor for its results, because we have not bothered to take a personal interest in either of them.

A simple intention rests in God while accomplishing all things. It takes account of particular ends in order to achieve them for Him: but it does not rest in them. Since a simple intention does not need to rest in any particular end, it has already reached the end as soon as the work is begun. For the end of a simple intention is to work in God and with Him—to sink deep roots into the soil of His will and to grow there in whatever weather He may bring.

A right intention is what we might call a "transient" intention: it is proper to the active life which is always moving on to something else. Our right intention passes from one particular end to another, from work to work, from day to day, from possibility to possibility. It reaches ahead into many plans. The works planned and done are all for the glory of God: but they stand ahead of us as milestones along a road with an invisible end. And God is always there at the end. He is always "future," even though He may be present. The spiritual life of a man of right intention is always more or less provisional. It is more possible than actual, for he always lives as if he had to finish just one more job before he could relax and look for a little contemplation.

Nevertheless, even in contemplative monasteries a "right" intention is more common than a really simple intention. Contemplatives, too, can live in a world where things to be done obscure the vision of Him for whom they are done. It makes no difference if these things to be done are within ourselves. Perhaps the confusion is only made more difficult by the fact that our right intention has nothing tangible to take hold of, and reaches out all day long for merits, sacrifices, degrees of virtue and of prayer. In fact, without a simple intention, a life of prayer tends to be not only difficult but even incomprehensible. For the aim of the contemplative life is not merely to enable a man to say prayers and make sacrifices with a right intention: it is to teach him to live in God.

18. Simple intention is a rare gift of God. Rare because it is poor. Poverty is a gift that few religious people really relish. They want their religion to make them at least spiritually rich, and if they renounce all things in this world, they want to lay hands not only on life everlasting but, above all, on the "hundredfold" promised to us even before we die.

Actually, that hundredfold is found in the beatitudes, the first of which is poverty.

Our intention cannot be completely simple unless it is completely poor. It seeks and desires nothing but the supreme poverty of having nothing but God. True, anyone with a grain of faith realizes that to have God and nothing else besides is to have everything in Him. But between the thought of such poverty and its actualization in our lives lies the desert of emptiness through which we must travel in order to find Him.

With a right intention, you quietly face the risk of losing the fruit of your work. With a simple intention you renounce the fruit before you even begin. You no longer even expect it. Only at this price can your work also become a prayer.

19. A simple intention is a perpetual death in Christ. It keeps our life hidden with Christ in God. It seeks its treasure nowhere except in heaven. It prefers what cannot be touched, counted, weighed, tasted, or seen. But it makes our inner being open out, at every moment, into the abyss of divine peace in which our life and actions have their roots.

A right intention aims only at right action.

But even in the midst of action, a simple intention, renouncing all things but God alone, seeks Him alone. The secret of simple intention is that it is content to seek God and does not insist on finding Him right away, knowing that in seeking Him it has already found Him. Right intention knows this too, but not by experience, and therefore it obscurely feels that seeking God is still not enough.

Simple intention is a divine medicine, a balm that soothes the powers of our soul wounded by inordinate self-expression. It heals our actions in their secret infirmity. It draws our strength to the hidden summit of our being, and bathes our spirit in the infinite mercy of God. It wounds our souls in order to heal them in Christ, for a simple intention manifests the presence and action of Christ in our hearts. It makes us His perfect instruments, and transforms us into His likeness, filling our whole lives with His gentleness and His strength and His purity and His prayer and His silence.

Whatever is offered to God with a right intention is acceptable to Him.

Whatever is offered to God with a simple intention is not only accepted by Him by reason of our good will, but is pleasing to Him in itself. It is a good and perfect work, performed entirely by His love. It draws its perfection not from our poor efforts alone but from His mercy which has made them rich. In giving the Lord the works of a right intention I can be sure that I am giving Him what is not bad. But in offering Him the works of a simple intention I am giving Him what is best. And beyond all that I can give Him or do for Him, I rest and take my joy in His glory.

5

The Word of the Cross

THE word of the Cross is foolishness, says St. Paul, to them that perish (I Corinthians 1 : 18). Yet among those to whom the Cross was folly and scandal were ascetics and religious men who had evolved a philosophy of suffering and who cultivated self-denial.

There is, therefore, much more in the word of the Cross than the acceptance of suffering or the practice of self-denial. The Cross is something positive. It is more than a death. The word of the Cross is foolishness to them that perish—but to them that are saved "it is the power of God" (I Corinthians 1 : 18).

2. The Christian must not only accept suffering: he must make it holy. Nothing so easily becomes unholy as suffering.

Merely accepted, suffering does nothing for our souls except, perhaps, to harden them. Endurance alone is no consecration. True asceticism is not a mere cult of fortitude. We can deny ourselves rigorously for the wrong reason and end up by pleasing ourselves mightily with our self-denial.

Suffering is consecrated to God by faith—not by faith in suffering, but by faith in God. To accept suffering stoically, to receive the burden of fatal, unavoidable, and incomprehensible necessity and to bear it strongly, is no consecration.

Some men believe in the power and the value of suffering. But their belief is an illusion. Suffering has no power and no value of its own.

It is valuable only as a test of faith. What if our faith fails in the test? Is it good to suffer, then? What if we enter into suffering with a strong faith in suffering, and then discover that suffering destroys us?

To believe in suffering is pride: but to suffer, believing in God, is humility. For pride may tell us that we are strong enough to suffer, that suffering is good for us because we are good. Humility tells us that suffering is an evil which we must always expect to find in our lives because of the evil that is in ourselves. But faith also knows that the mercy of God is given to those who seek Him in suffering, and that by His grace we can overcome evil with good. Suffering, then, becomes good by accident, by the good that it enables us to receive more abundantly from the mercy of God. It does not make us good by itself, but it enables us to make ourselves better than we are. Thus, what we consecrate to God in suffering is not our suffering but our *selves*.

3. Only the sufferings of Christ are valuable in the sight of God, who hates evil, and to Him they are valuable chiefly as a sign. The death of Jesus on the Cross has an infinite meaning and value not because it is a death, but because it is the death of the Son of God. The Cross of Christ says nothing of the power of suffering or of death. It speaks only of the power of Him who overcame both suffering and death by rising from the grave.

The wounds that evil stamped upon the flesh of Christ are to be worshipped as holy not because they are wounds, but because they are *His* wounds. Nor would we worship them if He had merely died of them, without rising again. For Jesus is not merely someone who once loved men enough to die for them. He is a man whose human nature subsists in God, so that He is a divine person. His love for us is the infinite love of God, which is stronger than all evil and cannot be touched by death.

Suffering, therefore, can only be consecrated to God by one who believes that Jesus is not dead. And it is of the very essence of Christianity to face suffering and death not because they are good, not because they have meaning, but because the Resurrection of Jesus has robbed them of their meaning.

4. The saint is not one who accepts suffering because he likes it, and confesses this preference before God and men in order to win a great reward. He is one who may well hate suffering as much as anybody else, but who so loves Christ, whom he does not see, that he will allow his love to be proved by any suffering. And he does this not because he thinks it is an achievement, but because the charity of Christ in his heart demands that it be done.

The saint is one so attuned to the spirit and heart of Christ that he is compelled to answer the demands of love by a love that matches that of Christ. This is for him a need so deep and so personal and so exacting that it becomes his whole destiny. The more he answers the secret action of Christ's love in his own heart, the more he comes to know that love's inexorable demands.

But the life of the Christian soul must always be a thing whole and simple and complete and incommunicable. The saints may seem to desire suffering in a universal and abstract way. Actually, the only sufferings anyone can validly desire are those precise, particular trials that are demanded of us in the designs of Divine Providence for our own lives.

Some men have been picked out to bear witness to Christ's love in lives overwhelmed by suffering. These have proclaimed that suffering was their vocation. But that should not lead us to believe that in order to be a saint one must go out for suffering in the same way that a college athlete goes out for football. No two men have to suffer exactly the same trials in exactly the same way. No one man is ever called to suffer merely for the sake of suffering.

What, after all, is more personal than suffering? The awful futility of our attempts to convey the reality of our sufferings to other people, and the tragic inadequacy of human sympathy, both prove how incommunicable a thing suffering really is.

When a man suffers, he is most alone. Therefore, it is in suffering that we are most tested as persons. How can we face the awful interior questioning? What shall we answer when we come to be examined by pain? Without God, we are no longer persons. We lose our manhood and our dignity. We become dumb animals under pain, happy if we can behave at least like quiet animals and die without too much commotion.

5. When suffering comes to put the question: "Who are you?" we must be able to answer distinctly, and give our own name. By that I mean we must express the very depths of what we are, what we have desired to be, what we have become. All these things are sifted out of us by

pain, and they are too often found to be in contradiction with one another. But if we have lived as Christians, our name and our work and our personality will fit the pattern stamped in our souls by the sacramental character we wear. We get a name in baptism. That is because the depths of our soul are stamped, by that holy sacrament, with a supernatural identification which will eternally tell us who we were meant to be. Our baptism, which drowns us in the death of Christ, summons upon us all the sufferings of our life: their mission is to help us work out the pattern of our identity received in the sacrament.

If, therefore, we desire to be what we are meant to be, and if we become what we are supposed to become, the interrogation of suffering will call forth from us both our own name and the name of Jesus. And we will find that we have begun to work out our destiny which is to be at once ourselves and Christ.

6. Suffering, and the consecration it demands, cannot be understood perfectly outside the context of baptism. For baptism, in giving us our identity, gives us a divine vocation to find ourselves in Christ. It gives us our identity in Christ. But both the grace and character of baptism give our soul a spiritual conformity to Christ *in His sufferings*. For baptism is the application to our souls of the Passion of Christ.

Baptism engrafts us into the mystical vine which is the body of Christ, and makes us live in His life and ripen like grapes on the trellis of His Cross. It brings us into the communion of the saints whose life flows from the Passion of Jesus. But every sacrament of union is also a sacrament of separation. In making us members of one another, baptism also more clearly distinguishes us, not only from those who do not live in Christ, but also and even especially from one another. For it gives us our personal, incommunicable vocation to reproduce in our own lives the life and sufferings and charity of Christ in a way unknown tc anyone else who has ever lived under the sun.

7. Suffering can only be perfectly consecrated to God, then, if it is seen as a fruit of baptism. It makes some sense only when it is plunged in the waters of the sacrament. Only these waters give it power to wash and purify. Only baptism sets out clearly *who* it is that must be formed and perfected by tribulation.

Suffering, therefore, must make sense to us not as a vague universal necessity, but as something demanded by our own personal destiny. When I see my trials not as the collision of my life with a blind machine called fate, but as the sacramental gift of Christ's love, given to me by God the Father along with my identity and my very name, then I can consecrate them and myself with them to God. For then I realize that my suffering is not my own. It is the Passion of Christ, stretching out its tendrils into my life in order to bear rich clusters of grapes, making my soul dizzy with the wine of Christ's love, and pouring that wine as strong as fire upon the whole world.

8. Useless and hateful in itself, suffering without faith is a curse.

A society whose whole idea is to eliminate suffering and bring all its members the greatest amount of comfort and pleasure is doomed to be destroyed. It does not understand that all evil is not necessarily to be avoided. Nor is suffering the only evil, as our world thinks.

If we consider suffering to be the greatest evil and pleasure the greatest good, we will live continually submerged in the only great evil that we ought to avoid without compromise: which is sin. Sometimes it is absolutely necessary to face suffering, which is a lesser evil, in order to avoid or to overcome the greatest evil, sin.

What is the difference between physical evil—suffering—and moral evil—sin? Physical evil has no power to penetrate beneath the surface of our being. It can touch our flesh, our mind, our sensibility. It cannot harm our spirit without the work of that other evil which is sin. It we suffer courageously, quietly, unselfishly, peacefully, the things that wreck our outer being only perfect us within, and make us, as we have seen, more truly ourselves because they enable us to fulfil our destiny in Christ. They are sent for this purpose, and when they come we should receive them with gratitude and joy.

Sin strikes at the very depth of our personality. It destroys the one reality on which our true character, identity, and happiness depend: our fundamental orientation to God. We are created to will what God wills, to know what He knows, to love what He loves. Sin is the will to do what God does not will, to know what He does not know, to love what He does not love. Therefore every sin is a sin against truth, a sin against obedience, and against love. But in all these three things sin proves itself to be a supreme injustice not only against God but, above all, against ourselves.

For what is the good of knowing what God does not know? To know what He does not know is to know what is not. And why love what He does not love? Is there any purpose in loving nothing: for He loves everything that is.

And our destiny is to love all things that He loves, just as He loves them. The will to love what is not is at the same time a refusal to love what is. And why should we destroy ourselves by willing what God does not will? To will against His will is to turn our will against ourselves. Our deepest spiritual need is for whatever thing God wills for us. To will something else is to deprive ourselves of life itself. So, when we sin, our spirit dies of starvation.

Physical evil is only to be regarded as a real evil in so far as it tends to foment sin in our souls. That is why a Christian must seek in every way possible to relieve the sufferings of others, and even take certain necessary steps to alleviate some sufferings of his own: because they are occasions of sin. It is true that we can also have compassion for others merely because suffering is an evil in its own right. This compassion is also good. But it does not really become charity unless it sees Christ in the one suffering and has mercy on him with the mercy of Christ. Jesus had pity on the multitudes not only because they were sheep without a shepherd, but also simply because they had no bread. Yet He did not feed them with miraculous loaves and fishes without thought for their place in His Father's Kingdom. Bodily works of mercy look beyond the flesh and into the spirit, and when they are integrally Christian they not only alleviate suffering but they bring grace: that is, they strike at sin.

9. Suffering is wasted if we suffer entirely alone. Those who do not know Christ, suffer alone. Their suffering is no communion. The awful solitude of suffering is not meant to seek communion in vain. But all communion is

denied to it except that which unites our spirit with God in the Passion of Jesus Christ.

What can human sympathy offer us in the loneliness of death? Flowers are an indecency in a death without God. They only serve to cover the body. The thing that has died has become a thing to be decorated and rejected. May its hopeless loneliness be forgotten and not remind us of our own!

How sad a thing is human love that ends with death: sadder when it pitiably tries to reach out to some futile communication with the dead. The poor little rice cakes at a pagan tomb! Sad, too, is the love that has no communion with those we love when they suffer. How miserable it is to have to stand in mute sorrow with nothing to say to those we love, when they are in great pain. It is a terrible confession that our love is not big enough to surmount suffering. Therefore we are desperately compelled to fight off suffering as long as we can, lest it come in the way and block off our love forever.

But a love that ends with either suffering or death is not worth the trouble it gives us. And if it must dread death and all suffering, it will inevitably bring us little joy and very much sorrow.

The Name and the Cross and the Blood of Jesus have changed all this. In His Passion, in the sacraments which bring His Passion into our lives, the helplessness of human love is transformed into a divine power which raises us above all evil. It has conquered everything. Such love knows no separation. It fears suffering no more than young crops fear the spring rain.

But the strength of such love and such communion is not found merely in a doctrine. The Christian has more than a philosophy of suffering. Sometimes, indeed, he may have no philosophy at all. His faith may be so inarticulate as to seem absurd. Nevertheless, he knows the peace of one who has conquered everything. Why is this? Because Christianity is Christ living in us, and Christ has conquered everything. Furthermore, He has united us to one another in Himself. We all live together in the power of His death which overcame death. We neither suffer alone nor conquer alone nor go off into eternity alone. In Him we are inseparable: therefore, we are free to be fruitfully alone whenever we please, because wherever we go, whatever we suffer, whatever happens to us, we are united with those we love in Him because we are united with Him.

His love is so much stronger than death that the death of a Christian is a kind of triumph. And although we rightly sorrow at the sensible separation from those we love (since we are also meant to love their human presence), yet we rejoice in their death because it proves to us the strength of our mutual love. The conviction in our hearts, the unshakable hope of communion with our dead in Christ, is always telling us that they live and that He lives and that we live. This is our great inheritance, which can only be increased by suffering well taken: this terrific grip of the divine life on our own souls, this grip of clean love that holds us so fast that it keeps us eternally free. This love, this life, this presence, is the witness that the spirit of Christ lives in us, and that we belong to Him, and that the Father has given us to Him, and no man shall snatch us out of His hand.

10. Heroism alone is useless, unless it be born of God. The fortitude given us in the charity of Christ is not compli-

cated by pride. First of all, divine strength is not usually given us until we are fully aware of our own weakness and know that the strength we receive is indeed received: and that it is a gift. Then, the fortitude that comes to us from God is His own strength, which is beyond comparison. And pride is born of comparison.

11. To know the Cross is not merely to know our own sufferings. For the Cross is the sign of salvation, and no man is saved by his own sufferings. To know the Cross is to know that we are saved by the sufferings of Christ; more, it is to know the love of Christ who underwent suffering and death in order to save us. It is, then, to know Christ. For to know His love is not merely to know the story of His love, but to experience in our spirit that we are loved by Him, and that in His love the Father manifests His own love for us, through His Spirit poured forth into our hearts. To know all this is to understand something of the Cross, that is: to know Christ. This explains the connection between suffering and contemplation. For contemplation is simply the penetration, by divine wisdom, into the mystery of God's love, in the Passion and Resurrection of Jesus Christ.

12. The holy were not holy because they were rejected by men, but because they were acceptable to God. Saints are not made saints merely by suffering.

The Lord did not create suffering. Pain and death came into the world with the fall of man. But after man had chosen suffering in preference to the joys of union with God, the Lord turned suffering itself into a way by which man could come to the perfect knowledge of God.

13. The effect of suffering upon us depends on what we love.

If we love ourselves selfishly, suffering is merely hateful. It has to be avoided at all costs. It brings out all the evil that is in us, so that the man who loves only himself will commit any sin and inflict any evil on others merely in order to avoid suffering himself.

Worse, if a man loves himself and learns that suffering is unavoidable, he may even come to take a perverse pleasure in suffering itself, showing that he loves and hates himself at the same time.

In any case, if we love ourselves, suffering inexorably brings out selfishness, and then, after making known what we are, drives us to make ourselves even worse than we are.

If we love others and suffer for them, even without a supernatural love for other men in God, suffering can give us a certain nobility and goodness. It brings out something fine in the nature of man, and gives glory to God who made man greater than suffering. But in the end a natural unselfishness cannot prevent suffering from destroying us along with all we love.

If we love God and love others in Him, we will be glad to let suffering destroy anything in us that God is pleased to let it destroy, because we know that all it destroys is unimportant. We will prefer to let the accidental trash of life be consumed by suffering in order that His glory may come out clean in everything we do.

If we love God, suffering does not matter. Christ in us, His love, His Passion in us: that is what we care about. Pain does not cease to be pain, but we can be glad of it because it enables Christ to suffer in us and give glory to

His Father by being greater, in our hearts, than suffering would ever be.

14. We have said that suffering has value in our lives only when it is consecrated to God. But consecration is a priestly act. Our sufferings then must be consecrated to God by His Church. She alone has power to drown our anguish in the Blood of Christ, for she alone possesses the infinite riches of His Passion and delegates men to exercise His priestly power. She alone has the seven sacraments by which the weakness and poverty of fallen man are transfigured in the death and Resurrection of Jesus.

It would not be enough for the Church to console us in our suffering, and her sacred rites are by no means meant to bring us only comfort.

The wisdom of the Church does, indeed, reach into the remotest corners of our human sensibility, for the Church is supremely human. But she is also divine. She anoints the souls of men with the Holy Spirit, stamping upon them a sacramental character which identifies them with the Crucified Christ, so that they can both suffer with His strength and compassionate the sufferings of others with His mercy.

Baptism makes all men share in the priesthood of Christ by sealing them with the sign of His death, which is our life. It gives them power to offer their sufferings, their good works, their acts of virtue, their courageous faith, their charitable mercy, their life itself to God, not as gifts of their own but as the sufferings, virtues, and merits of Jesus.

But baptism demands to be completed by the Eucharist. The Church has placed this most perfect sacrifice in the hands of her anointed priests, in whom Christ gathers the sufferings and sorrows and good works and joys of mankind to His heart and offers them all to the Father in the renewed sacrifice of His own body and blood.

15. When is suffering useless? When it only turns us in upon ourselves, when it only makes us sorry for ourselves, when it changes love into hatred, when it reduces all things to fear. Useless suffering cannot be consecrated to God because its fruitlessness is rooted in sin. Sin and useless suffering increase together. They encourage one another's growth, and the more suffering leads to sin, the more sin robs suffering of its capacity for fruitful consecration.

But the grace of Christ is constantly working miracles to turn useless suffering into something fruitful after all. How? By suddenly stanching the wound of sin. As soon as our life stops bleeding out of us in sin, suffering begins to have creative possibilities. But until we turn our wills to God, suffering leads nowhere, except to our own destruction.

16. The great duty of the religious soul is to suffer in silence. Too many men think they can become holy by talking about their trials. The awful fuss we sometimes make over the little unavoidable tribulations of life robs them of their fruitfulness. It turns them into occasions for self-pity or self-display, and consequently makes them useless.

Be careful of talking about what you suffer, for fear that you may sin. Job's friends sinned by the pious sententiousness of their explanation of suffering: and they sinned in giving Job a superficial explanation. Sometimes no explanation is sufficient to account for suffering. The only decent thing is silence—and the sacraments. The Church is very humble and very reserved in her treatment of suffering. She is never sanctimonious or patronizing. She is never sentimental. She knows what suffering is.

17. In order to face suffering in peace:

Suffer without imposing on others a theory of suffering, without weaving a new philosophy of life from your own material pain, without proclaiming yourself a martyr, without counting out the price of your courage, without disdaining sympathy and without seeking too much of it.

We must be sincere in our sufferings as in anything else. We must recognize at once our weakness and our pain, but we do not need to advertise them. It is well to realize that we are perhaps unable to suffer in grand style, but we must still accept our weakness with a kind of heroism. It is always difficult to suffer fruitfully and well, and the difficulty is all the greater when we have no human resources to help us. It is well, also, not to tempt God in our sufferings, not to extend ourselves, by pride, into an area where we cannot endure.

We must face the fact that it is much harder to stand the long monotony of slight suffering than a passing onslaught of intense pain. In either case what is hard is our own poverty, and the spectacle of our own selves reduced more and more to nothing, wasting away in our own estimation and in that of our friends.

We must be willing to accept also the bitter truth that, in the end, we may have to become a burden to those who love us. But it is necessary that we face this also. The full acceptance of our abjection and uselessness is the virtue that can make us and others rich in the grace of God. It takes heroic charity and humility to let others sustain us when we are absolutely incapable of sustaining ourselves.

We cannot suffer well unless we see Christ everywhere —both in suffering and in the charity of those who come to the aid of our affliction.

18. In order to give glory to God and overcome suffering with the charity of Christ:

Suffer without reflection, without hate, suffer with no hope of revenge or compensation, suffer without being impatient for the end of suffering.

Neither the beginning of suffering is important nor its ending. Neither the source of suffering is important nor its explanation, provided it be God's will. But we know that He does not will useless, that is to say sinful, suffering. Therefore in order to give Him glory we must be quiet and humble and poor in all that we suffer, so as not to add to our sufferings the burden of a useless and exaggerated sensibility.

In order to suffer without dwelling on our own affliction, we must think about a greater affliction, and turn to Christ on the Cross. In order to suffer without hate we must drive out bitterness from our heart by loving Jesus. In order to suffer without hope of compensation, we should find all our peace in the conviction of our union with Jesus. These things are not a matter of ascetic technique but of simple faith: they mean nothing without prayer, without desire, without the acceptance of God's will.

In the end, we must seek more than a passive acceptance of whatever comes to us from Him, we must desire and seek in all things the positive fulfilment of His will. We must suffer with gratitude, glad of a chance to do His will. And we must find, in this fulfilment, a communion with Jesus, who said: "With desire have I desired to eat this Pasch with you before I suffer" (Luke 22 : 15).

6

Asceticism and Sacrifice

If my soul silences my flesh by an act of violence, my flesh will take revenge on the soul, secretly infecting it with a spirit of revenge. Bitterness and bad temper are the flowers of an asceticism that has punished only the body. For the spirit is above the flesh, but not completely independent of the flesh. It reaps in itself what it sows in its own flesh. If the spirit is weak with the flesh, it will find in the flesh the image and accusation of its own weakness. But if the spirit is violent with the flesh it will suffer, from the flesh, the rebound of its own violence. The false ascetic begins by being cruel to everybody because he is cruel to himself. But he ends by being cruel to everybody but himself.

2. There is only one true asceticism: that which is guided not by our own spirit but by the Spirit of God. The spirit of man must first subject itself to grace and then it can bring the flesh in subjection both to grace and to itself. "If by the Spirit you mortify the deeds of the flesh, you shall live" (Romans 8 : 13).

But grace is charitable, merciful, kind, does not seek its own interests. Grace inspires us with no desire except to do the will of God, no matter what His will may be, no matter whether it be pleasing or unpleasant to our own nature.

Those, then, who put their passions to death not with

the poison of their own ambition but with the clean blade of the will of God will live in the silence of true interior peace, for their lives are hidden with Christ in God. Such is the meek "violence" of those who take Heaven by storm.

3. The spiritual life is not a mere negation of matter. When the New Testament speaks of "the flesh" as our enemy, it takes the flesh in a special sense. When Christ said: "The flesh profiteth nothing" (John 6 : 64), He was speaking of flesh without spirit, flesh living for its own ends, not only in sensual but even in spiritual things.

It is one thing to live *in* the flesh, and quite another to live *according to* the flesh. In the second case, one acquires that "prudence of the flesh which is opposed to God" because it makes the flesh an end in itself. But as long as we are on this earth our vocation demands that we live spiritually while still "in the flesh."

Our whole being, both body and soul, is to be spiritualized and elevated by grace. The Word who was made flesh and dwelt among us, who gave us His flesh to be our spiritual food, who sits at the right hand of God in a body full of divine glory, and who will one day raise our bodies also from the dead, did not mean us to despise the body or take it lightly when He told us to deny ourselves. We must indeed control the flesh, we must "chastise it and bring it into subjection," but this chastisement is as much for the body's benefit as for the soul's. For the good of the body is not found in the body alone but in the good of the whole person.

4. The spiritual man, who lives as a son of God, seeks the principle of his life above the flesh and above human nature itself. "As many as received Him, He gave them the

power to become the sons of God, to them that believe in His name. Who are born not of blood, nor of the will of the flesh, nor of the will of man, but of God" (John 1 : 12–13). God Himself, then, is the source of the spiritual life. But He communicates His life and His Spirit to men, made of body and soul. It is not His plan to lure the soul out of the body, but to sanctify the two together, divinizing the whole man so that the Christian can say: "I live, now not I, but Christ liveth in me. And that I live now in the flesh, I live in the faith of the Son of God who loved me" (Galatians 2 : 20). "That the justification of the law might be fulfilled in us who walk not according to the flesh but according to the spirit. . . . You are not in the flesh but in the Spirit if so be that the Spirit of God dwell in you" (Romans 8 : 4, 9).

5. We cannot become saints merely by trying to run away from material things. To have a spiritual life is to have a life that is spiritual in all its wholeness—a life in which the actions of the body are holy because of the soul, and the soul is holy because of God dwelling and acting in it. When we live such a life, the actions of our body are directed to God by God Himself and give Him glory, and at the same time they help to sanctify the soul.

The saint, therefore, is sanctified not only by fasting when he should fast but also by eating when he should eat. He is not only sanctified by his prayers in the darkness of the night, but by the sleep that he takes in obedience to God, who made us what we are. Not only His solitude contributes to his union with God, but also his supernatural love for his friends and his relatives and those with whom he lives and works.

God, in the same infinite act of will, wills the good of all beings and the good of each individual thing: for all lesser goods coincide in the one perfect good which is His love for them. Consequently it is clear that some men will become saints by a celibate life, but many more will become saints as married men, since it is necessary that there be more married men than celibates in the world. How then can we imagine that the cloister is the only place in which men can become saints? Now the life of the body seems to receive less consideration in the cloister than it does in secular life. But it is clear that married life, for its success, presupposes the capacity for a deeply human love which ought to be spiritual and physical at the same time. The existence of a sacrament of matrimony shows that the Church neither considers the body evil nor repugnant, but that the "flesh" spiritualized by prayer and the Holy Ghost, yet remaining completely physical, can come to play an important part in our sanctification.

6. It gives great glory to God for a person to live in this world using and appreciating the good things of life without care, without anxiety, and without inordinate passion. In order to know and love God through His gifts, we have to use them as if we used them not (I Corinthians 7: 31)—and yet we have to *use* them. For to use things as if we used them not means to use them without selfishness, without fear, without afterthought, and with perfect gratitude and confidence and love of God. All inordinate concern over the material side of life was reproved by Christ when He said: "What one of you, by taking thought, can add to his stature one cubit?" (Matthew 6: 27). But we cannot use created things without anxiety unless we

are detached from them. At the same time, we become detached from them by using them sparingly—and yet without anxiety.

The tremulous scrupulosity of those who are obsessed with pleasures they love and fear narrows their souls and makes it impossible for them to get away from their own flesh. They have tried to become spiritual by worrying about the flesh, and as a result they are haunted by it. They have ended in the flesh because they began in it, and the fruit of their anxious asceticism is that they "use things not," but do so as if they used them. In their very self-denial they defile themselves with what they pretend to avoid. They do not have the pleasure they seek, but they taste the bitter discouragement, the feeling of guilt which they would like to escape. This is not the way of the spirit. For when our intention is directed to God, our very use of material things sanctifies both them and us, provided we use them without selfishness and without presumption, glad to receive them from Him who loves us and whose love is all we desire.

7. Our self-denial is sterile and absurd if we practise it for the wrong reasons or, worse still, without any valid reason at all. Therefore, although it is true that we must deny ourselves in order to come to a true knowledge of God, we must also have some knowledge of God and of our relationship with Him in order to deny ourselves intelligently.

In order to be intelligent, our self-denial must first of all be humble. Otherwise it is a contradiction in terms. If we deny ourselves in order to think ourselves better than other men our self-denial is only self-gratification.

But our renunciation must be more than intelligent and humble. It must also be supernatural. It must be ordered not merely to our own moral perfection or to the good of the society we live in, but to God. Nothing comes to God but what comes from God, and our self-denial cannot be supernatural unless it be guided by the grace of the Holy Spirit. The light of His grace teaches us the distinction between what is good and evil in ourselves, what is from God and what is from ourselves, what is acceptable to God and what merely flatters our own self-esteem. But the Holy Spirit also teaches us the difference between asceticism and sacrifice, and shows us that for a Christian asceticism is not enough.

Asceticism is content systematically to mortify and control our nature. Sacrifice does something more: it offers our nature and all its faculties to God. A self-denial that is truly supernatural must aspire to offer God what we have renounced ourselves. The perfection of Christian renunciation is the total offering of ourselves to God in union with the sacrifice of Christ.

The meaning of this sacrifice of ourselves is that we renounce the dominion of our own acts and of our own life and of our own death into the hands of God so that we do all things not for ourselves or according to our own will and our own desires, but for God, and according to His will.

The spirit of Christian sacrifice is well described in these lines of St. Paul: "None of us liveth to himself, and no man dieth to himself. For whether we live, we live unto the Lord, and whether we die we die unto the Lord. Therefore whether we live or die, we are in the Lord. For to this end Christ died and rose again; that He might be the Lord of the dead and of the living" (Romans 14 : 7–9).

To offer this sacrifice perfectly we must practise asceticism, without which we cannot gain enough control over our hearts and their passions to reach such a degree of indifference to life and death. But here again the Holy Spirit teaches us that indifference is not enough. We must indeed become indifferent to the things we have renounced: but this indifference should be the effect of love for God, in whose honour we renounce them. More than an effect, it is really only an aspect of that love.

Because we love God alone, beyond and above all things, and because our love shows us that He infinitely exceeds the goodness of them all, we become indifferent to all that is not God. But at the same time our love enables us to find, in God Himself, the goodness and the reality of all the things we have renounced for His sake. We then see Him whom we love in the very things we have renounced, and find them again in Him. Although the grace of the Holy Spirit teaches us to use created things "as if we used them not"—that is to say, with detachment and indifference—it does not make us indifferent to the value of the things in themselves. On the contrary, it is only when we are detached from created things that we can begin to value them as we really should. It is only when we are "indifferent" to them that we can really begin to love them. The indifference of which I speak must, therefore, be an indifference not to things themselves but to their effects in our own lives.

The man who loves himself more than God, loves things and persons for the good he himself can get out of them. His selfish love tends to destroy them, to consume them, to absorb them into his own being. His love of them is only one aspect of his own selfishness. It is only a kind of prejudice in his own favour. Such a man is by no means indifferent to the impact of things, persons, and events on his own life. But he is really detached from the good of things and persons themselves, considered quite apart from any good of his own. With respect to the good he gets out of them, he is neither detached nor unconcerned. But with respect to their own good he is completely indifferent.

The man who loves God more than himself is also able to love persons and things for the good that they possess in God. That is the same as saying he loves the glory they give to God: for that glory is the reflection of God in the goodness He has given to His creatures. Such a man is indifferent to the impact of things in his own life. He considers things only in relation to God's glory and God's will. As far as his own temporal advantage and satisfaction go, he is detached and unconcerned. But he is no more indifferent to the value of things in themselves than he is indifferent to God. He loves them in the same act with which he loves God. That is: he loves them in the act by which he has renounced them. And in that love by renouncing them he has regained them on a higher level.

8. To say that Christian renunciation must be ordered to God is to say that it must bear fruit in a deep life of prayer and then in works of active charity. Christian renunciation is not a matter of technical self-denial, beginning and ending within the narrow limits of our own soul. It is the first movement of a liberty which escapes the boundaries of all that is finite and natural and contingent, enters into a contact of charity with the infinite goodness of God, and then goes forth from God to reach all that He loves.

Christian self-denial is only the beginning of a divine fulfilment. It is inseparable from the inward conversion of our whole being from ourselves to God. It is the denial of our unfulfilment, the renunciation of our own poverty, that we may be able to plunge freely into the plenitude and the riches of God and of His creation without looking back upon our own nothingness.

Self-denial delivers us from the passions and from selfishness. It delivers us from a superstitious attachment to our own ego as if it were a god. It delivers us from the "flesh" in the technical New Testament sense, but it does not deliver us from the body. It is no escape from matter or from the senses, nor is it meant to be. It is the first step toward a transformation of our entire being in which, according to the plan of God, even our bodies will live in the light of His divine glory and be transformed in Him together with our souls.

9. Nothing that we consider evil can be offered to God in sacrifice. Therefore, to renounce life in disgust is no sacrifice. We give Him the best we have, in order to declare that He is infinitely better. We give Him all that we prize, in order to assure Him that He is more to us than our "all." One of the chief tasks of Christian asceticism is to make our life and our body valuable enough to be offered to God in sacrifice.

Our asceticism is not supposed to make us weary of a life that is vile. It is not supposed to make our bodies, which are good, appear to us to be evil. It is not supposed to make us odious to ourselves. An asceticism that makes all pleasure seem gross and disgusting and all the activities of the flesh abominable is a perversion of the nature which

God made good and which even sin has not succeeded in rendering totally vile.

The real purpose of asceticism is to disclose the difference between the evil use of created things, which is sin, and their good use, which is virtue. It is true that our self-denial teaches us to realize that sin, which appears to be good from a certain point of view, is really evil. But self-denial should not make us forget the essential distinction between sin, which is a negation, and pleasure, which is a positive good. In fact, it should make that distinction clearly known. True asceticism shows us that there is no necessary connection between sin and pleasure: that there can be sins that seek no pleasure, and other sins that find none.

Pleasure, which is good, has more to do with virtue than it has with sin. The virtue that is sufficiently resolute to pay the price of self-denial will eventually taste greater pleasure in the things it has renounced than could ever be enjoyed by the sinner who clings to those same things as desperately as if they were his god.

We must, therefore, gain possession of ourselves, by asceticism, in order that we may be able to give ourselves to God. No inspiration of the Spirit of God will ever move us to cast off the body as if it were evil, or to destroy its faculties as if they were the implacable enemies of God and could never be educated to obey His grace. He who made our flesh and gave it to our spirit as its servant and companion will not be pleased by a sacrifice in which the flesh is murdered by the spirit and returned to Him in ruin.

Yet someone will say that many of the saints did, in fact, walk into God's Heaven upon the ruins of their body. If they did so, and if they were really saints, it was not because their flesh was destroyed by their own spirit but because the love of God, which possessed them, led them into a situation in which the renunciation of health or of life itself was necessary for the sake of some greater good. The sacrifice was both justified and holy. They did not consider their own flesh evil, and destroy it for being so. They knew that it was good and that it came from God, but they knew that charity itself is a greater good than life, and that man has no greater love than that he lay down his life for his friend. It was for the sake of love, for the sake of other men, or for God's truth, that they sacrificed their bodies. For no man can become a saint merely by hating himself. Sanctity is the exact opposite of suicide.

10. There is no such thing as a sacrifice of ourselves that is merely self-destruction. We sacrifice ourselves to God by the spiritualization of our whole being through obedience to His grace. The only sacrifice He accepts is the purity of our love. Any renunciation that helps us to love God more is good and useful. A renunciation that may be noble in itself is useless for us if God does not will us to make it.

In order to spiritualize our lives and make them pleasing to God, we must become quiet. The peace of a soul that is detached from all things and from itself is the sign that our sacrifice is truly acceptable to God.

Bodily agitation agitates the soul. But we cannot tranquillize our spirit by forcing a violent immobility upon the flesh and its five senses. The body must be governed in such a way that it works peacefully, so that its action does not disturb the soul.

Peace of soul does not, therefore, depend on physical inactivity. On the contrary, there are some people who are perfectly capable of tasting true spiritual peace in an active life but who would go crazy if they had to keep themselves still in absolute solitude and silence for any length of time.

It is for each one to find out for himself the kind of work and environment in which he can best lead a spiritual life. If it is possible to find such conditions, and if he is able to take advantage of them, he should do so. But what a hopeless thing the spiritual life would be if it could only be lived under ideal conditions! Such conditions have never been within the reach of most men, and were never more inaccessible than in our modern world. Everything in modern city life is calculated to keep man from entering into himself and thinking about spiritual things. Even with the best of intentions a spiritual man finds himself exhausted and deadened and debased by the constant noise of machines and loudspeakers, the dead air and the glaring lights of offices and shops, the everlasting suggestions of advertising and propaganda.

The whole mechanism of modern life is geared for a flight from God and from the spirit into the wilderness of neurosis. Even our monasteries are not free from the smell and clatter of our world.

Bodily agitation, then, is an enemy to the spirit. And by agitation I do not necessarily mean exercise or movement. There is all the difference in the world between agitation and work.

Work occupies the body and the mind and is necessary for the health of the spirit. Work can help us to pray and be recollected if we work properly. Agitation, however, destroys the spiritual usefulness of work and even tends to frustrate its physical and social purpose. Agitation is the useless and ill-directed action of the body. It expresses the inner confusion of a soul without peace. Work brings peace to the soul that has a semblance of order and spiritual understanding. It helps the soul to focus upon its spiritual aims and to achieve them. But the whole reason for agitation is to hide the soul from itself, to camouflage its interior conflicts and their purposelessness, and to induce a false feeling that "we are getting somewhere." Agitation—a condition of spirit that is quite normal in the world of business—is the fruit of tension in a spirit that is turning dizzily from one stimulus to another and trying to react to fifteen different appeals at the same time. Under the surface of agitation, and furnishing it with its monstrous and inexhaustible drive, is the force of fear or elemental greed for money, or pleasure, or power. The more complex a man's passions, the more complex his agitation. All this is the death of the interior life. Occasional churchgoing and the recitation of hasty prayers have no power to cleanse this purulent wound.

No matter what our aims may be, no matter how spiritual, no matter how intent we think we are upon the glory of God and His Kingdom, greed and passion enter into our work and turn it into agitation as soon as our intention ceases to be pure. And who can swear that his intentions are pure, even down to the subconscious depths of his will, where ancient selfish motives move comfortably like forgotten sea monsters in waters where they are never seen!

In order to defend ourselves against agitation, we must be detached not only from the immediate results of our work—and this detachment is difficult and rare—but from the whole complex of aims that govern our earthly lives. We have to be detached from health and security, from pleasures and possessions, from people and places and conditions and things. We have to be indifferent to life itself, in the Gospel sense, living like the lilies of the field, seeking first the Kingdom of Heaven and trusting that all our material needs will be taken care of into the bargain. How many of us can say, with any assurance, that we have even begun to live like this?

Lacking this detachment, we are subject to a thousand fears corresponding to our thousand anxious desires. Everything we love is uncertain: when we are seeking it, we fear we may not get it. When we have obtained it, we fear even more that it may be lost. Every threat to our security turns our work into agitation. Even a word, even the imagined thought we place in the mind of another, suspecting him of suspecting us—these are enough to turn our day into a millrace of confusion and anxiety and haste and who knows what other worse things besides!

We must, first of all, gain a supernatural perspective, see all things in the light of faith, and then we will begin the long, arduous labour of getting rid of all our irrational fears and desires. Only a relatively spiritual man is able even to begin this work with enough delicacy to avoid becoming agitated in his very asceticism!

11. It is just as easy to become attached to an ascetic technique as to anything else under the sun. But that does not mean we must renounce all thought of being systematic in our self-denial. Let us only be careful to remember that systems are not ends in themselves. They are means to an end. Their proximate end is to bring peace and calm to a detached spirit, to liberate the spirit from its passions, so that we can respond more readily to reason and to divine

grace. The ultimate end of all techniques, when they are used in the Christian context, is charity and union with God.

Discipline is not effective unless it is systematic, for the lack of system usually betrays a lack of purpose. Good habits are only developed by repeated acts, and we cannot discipline ourselves to do the same thing over again with any degree of intelligence unless we go about it systematically. It is necessary, above all in the beginning of our spiritual life, to do certain things at fixed times: fasting on certain days, prayer and meditation at definite hours of the day, regular examinations of conscience, regularity in frequenting the sacraments, systematic application to our duties of state, particular attention to virtues which are most necessary for us.

To desire a spiritual life is, thus, to desire discipline. Otherwise our desire is an illusion. It is true that discipline is supposed to bring us, eventually, to spiritual liberty. Therefore our asceticism should make us spiritually flexible, not rigid, for rigidity and liberty never agree. But our discipline must, nevertheless, have a certain element of severity about it. Otherwise it will never set us free from the passions. If we are not strict with ourselves, our own flesh will soon deceive us. If we do not command ourselves severely to pray and do penance at certain definite times, and make up our mind to keep our resolutions in spite of notable inconvenience and difficulty, we will quickly be deluded by our own excuses and let ourselves be led away by weakness and caprice.

12. It is very helpful to have a spiritual director who will guide our efforts in self-discipline, and although direction is not absolutely necessary, in theory, for a sound spiritual

life, there are, nevertheless, in practice many men who will never get anywhere without it. Besides the valuable instruction a good director can give us, we also need his encouragement and his corrections. It is much easier to persevere in our penance, meditation, and prayer if we have someone to remind us of the resolutions we have begun to forget. Spiritual direction will protect us, in some measure, against our own instability. The function of the director is to orientate our discipline toward spiritual freedom. It takes a good director to do this, and good directors are rare.

13. Asceticism is utterly useless if it turns us into freaks. The cornerstone of all asceticism is humility, and Christian humility is first of all a matter of supernatural common sense. It teaches us to take ourselves as we are, instead of pretending (as pride would have us imagine) that we are something better than we are. If we really know ourselves we quietly take our proper place in the order designed by God. And so supernatural humility adds much to our human dignity by integrating us in the society of other men and placing us in our right relation to them and to God. Pride makes us artificial, and humility makes us real.

St. Paul teaches (II Thessalonians 3) that Christian humility and asceticism should normally help us to lead quite ordinary lives, peacefully earning our bread and working from day to day in a world that will pass away. Work and a supernatural acceptance of ordinary life are seen by the Apostle as a protection against the restless agitation of false mysticism. The Christian has rejected all the values of the world. He does not set his heart on temporal security and happiness. But it does not follow that he cannot continue to live in the world, or be happy in time. He works and lives in simplicity, with more joy and greater security than other men, because he does not look for any special fulfilment in this life. He avoids the futile agitation that surrounds the pursuit of purely temporal ends. He lives in peace amid the vanity of transient things. Nor does he merely despise their vanity: for behind the shadow he sees the substance, and creatures speak to him of joy in their Creator. It is supreme humility to see that ordinary life, embraced with perfect faith, can be more saintly and more supernatural than a spectacular ascetical career. Such humility dares to be ordinary, and that is something beyond the reach of spiritual pride. Pride always longs to be unusual. Humility not so. Humility finds all its peace in hope, knowing that Christ must come again to elevate and transfigure ordinary things and fill them with His glory.

14. God is more glorified by a man who uses the good things of this life in simplicity and with gratitude than by the nervous asceticism of someone who is agitated about every detail of his self-denial. The former uses good things and thinks of God. The latter is afraid of good things, and consequently cannot use them properly. He is terrified of the pleasure God has put in things, and in his terror thinks only of himself. He imagines God has placed all the good things of the world before him like a bait in a trap. He worries at all times about his own "perfection." His struggle for perfection becomes a kind of battle of wits with the Creator who made all things good. The very goodness of creatures becomes a threat to the purity of this virtuous one, who would like to abstain from everything

But he cannot. He is human, like the rest of men, and must make use like them of food and drink and sleep. Like them he must see the sky, and love, in spite of himself, the light of the sun! Every feeling of pleasure fills him with a sense of guilt. It has besmirched his own adored perfection. Strange that people like this should enter monasteries, which have no other reason for existing than the love of God!

15. All nature is meant to make us think of paradise. Woods, fields, valleys, hills, the rivers and the sea, the clouds travelling across the sky, light and darkness, sun and stars, remind us that the world was first created as a paradise for the first Adam, and that in spite of his sin and ours, it will once again become a paradise when we are all risen from death in the second Adam. Heaven is even now mirrored in created things. All God's creatures invite us to forget our vain cares and enter into our own hearts, which God Himself has made to be His paradise and our own. If we have God dwelling within us, making our souls His paradise, then the world around us can also become for us what it was meant to be for Adam—his paradise. But if we seek paradise outside ourselves, we cannot have paradise in our hearts. If we have no peace within ourselves, we have no peace with what is all around us. Only the man who is free from attachment finds that creatures have become his friends. As long as he is attached to them, they speak to him only of his own desires. Or they remind him of his sins. When he is selfish, they serve his selfishness. When he is pure, they speak to him of God.

16. If we are not grateful to God, we cannot taste the joy of finding Him in His creation. To be ungrateful is to

admit that we do not know Him, and that we love His creatures not for His sake but for our own. Unless we are grateful for our own existence, we do not know who we are, and we have not yet discovered what it really means to be and to live. No matter how high an estimate we may have of our own goodness, that estimate is too low unless we realize that all we have comes to us from God.

The only value of our life is that it is a gift of God.

Gratitude shows reverence to God in the way it makes use of His gifts.

7

Being and Doing

We are warmed by fire, not by the smoke of the fire. We are carried over the sea by a ship, not by the wake of a ship. So, too, what we are is to be sought in the invisible depths of our own being, not in our outward reflection in our own acts. We must find our real selves not in the froth stirred up by the impact of our being upon the beings around us, but in our own soul which is the principle of all our acts.

But my soul is hidden and invisible. I cannot see it directly, for it is hidden even from myself. Nor can I see my own eyes. They are too close to me for me to see them. They are not meant to see themselves. I know I have eyes when I see other things with them.

I can see my eyes in a mirror. My soul can also reflect itself in the mirror of its own activity. But what is seen in the mirror is only the reflection of who I am, not my true being. The mirror of words and actions only partly manifests my being.

The words and acts that proceed from myself and are accomplished outside myself are dead things compared with the hidden life from which they spring. These acts are transient and superficial. They are quickly gone, even though their effects may persist for a little while. But the soul itself remains. Much depends on how the soul sees itself in the mirror of its own activity.

2. My soul does not find itself unless it acts. Therefore it must act. Stagnation and inactivity bring spiritual death. But my soul must not project itself entirely into the outward effects of its activity. I do not need to *see* myself, I merely need to *be* myself. I must think and act like a living being, but I must not plunge my whole self into what I think and do, or seek always to find myself in the work I have done. The soul that projects itself entirely into activity, and seeks itself outside itself in the work of its own will is like a madman who sleeps on the pavement in front of his house instead of living inside where it is quiet and warm. The soul that throws itself outdoors in order to find itself in the effects of its own work is like a fire that has no desire to burn but seeks only to go up in smoke.

The reason why men are so anxious to see themselves, instead of being content to be themselves, is that they do not really believe in their own existence. And they do not fully believe that they exist because they do not believe in God. This is equally true of those who say they believe in God (without actually putting their faith into practice) and of those who do not even pretend to have any faith.

In either case, the loss of faith has involved at the same time a complete loss of all sense of reality. Being means nothing to those who hate and fear what they themselves are. Therefore they cannot have peace in their own reality (which reflects the reality of God). They must struggle to escape their true being, and verify a false existence by constantly viewing what they themselves do. They have to keep looking in the mirror for reassurance. What do they expect to see? Not themselves! They are hoping for some sign that they have become the god they hope to become by means of their own frantic activity—invulnerable, all-powerful, infinitely wise, unbearably beautiful, unable to die!

When a man constantly looks and looks at himself in the mirror of his own acts, his spiritual double vision splits him into two people. And if he strains his eyes hard enough, he forgets which one is real. In fact, reality is no longer found either in himself or in his shadow. The substance has gone out of itself into the shadow, and he has become two shadows instead of one real person.

Then the battle begins. Whereas one shadow was meant to praise the other, now one shadow accuses the other. The activity that was meant to exalt him reproaches and condemns him. It is never real enough. Never active enough. The less he is able to *be* the more he has to *do*. He becomes his own slave driver—a shadow whipping a shadow to death, because it cannot produce reality, infinitely substantial reality, out of his own nonentity.

Then comes fear. The shadow becomes afraid of the shadow. He who "is not" becomes terrified at the things he cannot do. Whereas for a while he had illusions of infinite power, miraculous sanctity (which he was able to guess at in the mirror of his virtuous actions), now it has all changed. Tidal waves of nonentity, of powerlessness, of hopelessness surge up within him at every action he attempts.

Then the shadow judges and hates the shadow who is not a god, and who can do absolutely nothing.

Self-contemplation leads to the most terrible despair: the despair of a god that hates himself to death. This is the ultimate perversion of man who was made in the image and likeness of the true God, who was made to love eternally and perfectly an infinite good—a good (note this well) which he was to find *dwelling within himself!*

In order to find God in ourselves, we must stop looking at ourselves, stop checking and verifying ourselves in the mirror of our own futility, and be content to *be* in Him and to do whatever He wills, according to our limitations, judging our acts not in the light of our own illusions, but in the light of His reality which is all around us in the things and people we live with.

3. All men seek peace first of all with themselves. That is necessary, because we do not naturally find rest even in our own being. We have to learn to commune with ourselves before we can communicate with other men and with God. A man who is not at peace with himself necessarily projects his interior fighting into the society of those he lives with, and spreads a contagion of conflict all around him. Even when he tries to do good to others his efforts are hopeless, since he does not know how to do good to himself. In moments of wildest idealism he may take it into his head to make other people happy: and in doing so he will overwhelm them with his own unhappiness. He seeks to find himself somehow in the work of making others happy. Therefore he throws himself into the work. As a result he gets out of the work all that he puts into it: his own confusion, his own disintegration, his own unhappiness.

It is useless to try to make peace with ourselves by being pleased with everything we have done. In order to settle down in the quiet of our own being we must learn to be detached from the results of our own activity. We must withdraw ourselves, to some extent, from effects that are beyond our control and be content with the good will and the work that are the quiet expression of our inner life. We must be content to live without watching ourselves live, to work without expecting an immediate reward, to love without an instantaneous satisfaction, and to exist without any special recognition.

It is only when we are detached from ourselves that we can be at peace with ourselves. We cannot find happiness in our work if we are always extending ourselves beyond ourselves and beyond the sphere of our work in order to find ourselves greater than we are.

Our Christian destiny is, in fact, a great one: but we cannot achieve greatness unless we lose all interest in being great. For our own idea of greatness is illusory, and if we pay too much attention to it we will be lured out of the peace and stability of the being God gave us, and seek to live in a myth we have created for ourselves. It is, therefore, a very great thing to be little, which is to say: to be ourselves. And when we are truly ourselves we lose most of the futile self-consciousness that keeps us constantly comparing ourselves with others in order to see how big we are.

4. The fact that our being necessarily demands to be expressed in action should not lead us to believe that as soon as we stop acting we cease to exist. We do not live merely in order to "do something"—no matter what. Activity is just one of the normal expressions of life, and the life it expresses is all the more perfect when it sustains itself with an ordered economy of action. This order demands a wise alternation of activity and rest. We do not live more fully merely by doing more, seeing more,

tasting more, and experiencing more than we ever have before. On the contrary, some of us need to discover that we will not begin to live more fully until we have the courage to do and see and taste and experience much less than usual.

A tourist may go through a museum with a Baedeker, looking conscientiously at everything important, and come out less alive than when he went in. He has looked at everything and seen nothing. He has done a great deal and it has only made him tired. If he had stopped for a moment to look at one picture he really liked and forgotten about all the others, he might console himself with the thought that he had not completely wasted his time. He would have discovered something not only outside himself but in himself. He would have become aware of a new level of being in himself and his life would have been increased by a new capacity for being and for doing.

Our being is not to be enriched merely by activity or experience as such. Everything depends on the *quality* of our acts and our experiences. A multitude of badly performed actions and of experiences only half-lived exhausts and depletes our being. By doing things badly we make ourselves less real. This growing unreality cannot help but make us unhappy and fill us with a sense of guilt. But the purity of our conscience has a natural proportion with the depth of our being and the quality of our acts: and when our activity is habitually disordered, our malformed conscience can think of nothing better to tell us than to multiply the *quantity* of our acts, without perfecting their quality. And so we go from bad to worse, exhaust ourselves, empty our whole life of all content, and fall into despair.

There are times, then, when in order to keep ourselves in existence at all we simply have to sit back for a while and do nothing. And for a man who has let himself be drawn completely out of himself by his activity, nothing is more difficult than to sit still and rest, doing nothing at all. The very act of resting is the hardest and most courageous act he can perform: and often it is quite beyond his power.

We must first recover the possession of our own being before we can act wisely or taste any experience in its human reality. As long as we are not in our own possession, all our activity is futile. If we let all our wine run out of the barrel and down the street, how will our thirst be quenched?

5. The value of our activity depends almost entirely on the humility to accept ourselves as we are. The reason why we do things so badly is that we are not content to do what we can.

We insist on doing what is not asked of us, because we want to taste the success that belongs to somebody else.

We never discover what it is like to make a success of our own work, because we do not want to undertake any work that is merely proportionate to our powers.

Who is willing to be satisfied with a job that expresses all his limitations? He will accept such work only as a "means of livelihood" while he waits to discover his "true vocation." The world is full of unsuccessful businessmen who still secretly believe they were meant to be artists or writers or actors in the movies.

6. The fruitfulness of our life depends in large measure on our ability to doubt our own words and to question the

value of our own work. The man who completely trusts his own estimate of himself is doomed to sterility. All he asks of any act he performs is that it be *his* act. If it is performed by him, it must be good. All words spoken by him must be infallible. The car he has just bought is the best for its price, for no other reason than that he is the one who has bought it. He seeks no other fruit than this, and therefore he generally gets no other.

If we believe ourselves in part, we may be right about ourselves. If we are completely taken in by our own disguise, we cannot help being wrong.

7. The measure of our being is not to be sought in the violence of our experiences. Turbulence of spirit is a sign of spiritual weakness. When delights spring out of our depths like leopards we have nothing to be proud of: our soul's life is in danger. For when we are strong we are always much greater than the things that happen to us, and the soul of a man who has found himself is like a deep sea in which there may be many fish: but they never come up out of the sea, and not one of them is big enough to trouble its placid surface. His "being" is far greater than anything he feels or does.

8. The deep secrecy of my own being is often hidden from me by my own estimate of what I am. My idea of what I am is falsified by my admiration for what I do. And my illusions about myself are bred by contagion from the illusions of other men. We all seek to imitate one another's imagined greatness.

If I do not know who I am, it is because I think I am the sort of person everyone around me wants to be. Perhaps

I have never asked myself whether I really wanted to become what everybody else seems to want to become. Perhaps if I only realized that I do not admire what everyone seems to admire, I would really begin to live after all. I would be liberated from the painful duty of saying what I really do not think and of acting in a way that betrays God's truth and the integrity of my own soul.

Why do we have to spend our lives striving to be something that we would never want to be, if we only knew what we wanted? Why do we waste our time doing things which, if we only stopped to think about them, are just the opposite of what we were made for?

We cannot be ourselves unless we know ourselves. But self-knowledge is impossible when thoughtless and automatic activity keeps our souls in confusion. In order to know ourselves it is not necessary to cease all activity in order to think about ourselves. That would be useless, and would probably do most of us a great deal of harm. But we have to cut down our activity to the point where we can think calmly and reasonably about our actions. We cannot begin to know ourselves until we can see the real reasons why we do the things we do, and we cannot be ourselves until our actions correspond to our intentions and our intentions are appropriate to our own situation. But that is enough. It is not necessary that we succeed in everything. A man can be perfect and still reap no fruit from his work, and it may happen that a man who is able to accomplish very little is much more of a person than another who seems to accomplish very much.

9. A man who fails well is greater than one who succeeds badly.

One who is content with what he has, and who accepts the fact that he inevitably misses very much in life, is far better off than one who has much more but who worries about all he may be missing. For we cannot make the best of what we are, if our hearts are always divided between what we are and what we are not.

The lower our estimate of ourselves and the lower our expectations, the greater chance we have of using what we have. If we do not know how poor we are we will never be able to appreciate what we actually have. But, above all, we must learn our own weakness in order to awaken to a new order of action and of being—and experience God Himself accomplishing in us the things we find impossible.

We cannot be happy if we expect to live all the time at the highest peak of intensity. Happiness is not a matter of intensity but of balance and order and rhythm and harmony.

Music is pleasing not only because of the sound but because of the silence that is in it: without the alternation of sound and silence there would be no rhythm. If we strive to be happy by filling all the silences of life with sound, productive by turning all life's leisure into work, and real by turning all our being into doing, we will only succeed in producing a hell on earth.

If we have no silence, God is not heard in our music. If we have no rest, God does not bless our work. If we twist our lives out of shape in order to fill every corner of them with action and experience, God will silently withdraw from our hearts and leave us empty.

Let us, therefore, learn to pass from one imperfect activity to another without worrying too much about what we are missing. It is true that we make many mistakes. But the biggest of them all is to be surprised at them: as if we had some hope of never making any.

Mistakes are part of our life, and not the least important part. If we are humble, and if we believe in the Providence of God, we will see that our mistakes are not merely a necessary evil, something we must lament and count as lost: they enter into the very structure of our existence. It is by making mistakes that we gain experience, not only for ourselves but for others. And though our experience prevents neither ourselves nor others from making the same mistake many times, the repeated experience still has a positive value.

10. We cannot avoid missing the point of almost everything we do. But what of it? Life is not a matter of getting something out of everything. Life itself is imperfect. All created beings begin to die as soon as they begin to live, and no one expects any one of them to become absolutely perfect, still less to stay that way. Each individual thing is only a sketch of the specific perfection planned for its kind. Why should we ask it to be anything more?

If we are too anxious to find absolute perfection in created things we cease to look for perfection where alone it can be found: in God. The secret of the imperfection of all things, of their inconstancy, their fragility, their falling into nothingness, is that they are only a shadowy expression of the one Being from whom they receive their being. If they were absolutely perfect and changeless in themselves, they would fail in their vocation, which is to give glory to God by their contingency.

It was the desire to "be as gods"—changelessly perfect in their own being—that led Adam and Eve to taste the fruit of the forbidden tree. What could be duller than an immutable man and an unchanging woman, eternally the same! As long as we are on earth our vocation is precisely to be imperfect, incomplete, insufficient in ourselves, changing, hapless, destitute, and weak, hastening toward the grave. But the power of God and His eternity and His peace and His completeness and His glory must somehow find their way into our lives, secretly, while we are here, in order that we may be found in Him eternally as He has meant us to be. And in Him, in our eternity, there will be no change in the sense of corruption, but there will be unending variety, newness of life, progression in His infinite depth. There, rest and action will not alternate, they will be one. Everything will be at once empty and full. But only if we have discovered how to combine emptiness and fullness, good will and indifferent results, mistakes and successes, work and rest, suffering and joy, in such a way that all things work together for our good and for the glory of God.

The relative perfection which we must attain to in this life if we are to live as sons of God is not the twenty-four-hour-a-day production of perfect acts of virtue, but a life from which practically all the obstacles to God's love have been removed or overcome.

One of the chief obstacles to this perfection of selfless charity is the selfish anxiety to get the most out of everything, to be a brilliant success in our own eyes and in the eyes of other men. We can only get rid of this anxiety by being content to miss something in almost everything we do. We cannot master everything, taste everything, understand everything, drain every experience to its last dregs. But if we have the courage to let almost everything else go, we will probably be able to retain the one thing necessary for us—whatever it may be. If we are too eager to have everything, we will almost certainly miss even the one thing we need.

Happiness consists in finding out precisely what the "one thing necessary" may be in our lives, and in gladly relinquishing all the rest. For then, by a divine paradox, we find that everything else is given us together with the one thing we needed.

8

Vocation

EACH one of us has some kind of vocation. We are all called by God to share in His life and in His Kingdom. Each one of us is called to a special place in the Kingdom. If we find that place, we will be happy. If we do not find it, we can never be completely happy. For each one of us, there is only one thing necessary: to fulfil our own destiny, according to God's will, to be what God wants us to be.

We must not imagine that we only discover this destiny by a game of hide-and-seek with Divine Providence. Our vocation is not a sphinx's riddle, which we must solve in one guess or else perish. Some people find, in the end, that they have made many wrong guesses and that their paradoxical vocation is to go through life guessing wrong. It takes them a long time to find out that they are happier that way.

In any case, our destiny is the work of two wills, not one. It is not an immutable fate, forced upon us without any choice of our own, by a divinity without heart.

Our vocation is not a supernatural lottery but the interaction of two freedoms, and, therefore, of two loves. It is hopeless to try to settle the problem of vocation outside the context of friendship and of love. We speak of Providence: that is a philosophical term. The Bible speaks of our Father in Heaven. Providence is, consequently, more than an institution, it is a person. More than a benevolent stranger, He is our Father. And even the term Father is too loose a metaphor to contain all the depths of the mystery: for He loves us more than we love ourselves, as if we were Himself. He loves us, moreover, with our own wills, with our own decisions. How can we understand the mystery of our union with God who is closer to us than we are to ourselves? It is His very closeness that makes it difficult for us to think of Him. He who is infinitely above us, infinitely different from ourselves, infinitely "other" from us, nevertheless dwells in our souls, watches over every movement of our life with as much love as if we were His own self. His love is at work bringing good out of all our mistakes and defeating even our sins.

In planning the course of our lives, we must remember the importance and the dignity of our own freedom. A man who fears to settle his future by a good act of his own free choice does not understand the love of God. For our freedom is a gift God has given us in order that He may be able to love us more perfectly, and be loved by us more perfectly in return.

2. Love is perfect in proportion to its freedom. It is free in proportion to its purity. We act most freely when we act purely in response to the love of God. But the purest love of God is not servile, not blind, not limited by fear. Pure charity is fully aware of the power of its own freedom. Perfectly confident of being loved by God, the soul that loves Him dares to make a choice of its own, knowing that its own choice will be acceptable to love.

At the same time pure love is prudent. It is enlightened

with a clear-sighted discretion. Trained in freedom, it knows how to avoid the selfishness that frustrates its action. It sees obstacles and avoids or overcomes them. It is keenly sensitive to the smallest signs of God's will and good pleasure in the circumstances of its own life, and its freedom is conditioned by the knowledge of all these. Therefore, in choosing what will please God, it takes account of all the slightest indications of His will. Yet if we add all these indications together, they seldom suffice to give us absolute certitude that God wills one thing to the exclusion of every other. He who loves us means by this to leave us room for our own freedom, so that we may dare to choose for ourselves, with no other certainty than that His love will be pleased by our intention to please Him.

3. Every man has a vocation to *be* someone: but he must understand clearly that in order to fulfil this vocation he can only be one person: himself.

Yet we have said that baptism gives us a sacramental character, defining our vocation in a very particular way since it tells us we must become ourselves in Christ. We must achieve our identity in Him, with whom we are already sacramentally identified by water and the Holy Spirit.

What does this mean? We must be ourselves by being Christ. For a man, to be is to live. A man only lives as a man when he knows truth and loves what he knows and acts according to what he loves. In this way he *becomes* the truth that he loves. So we "become" Christ by knowledge and by love.

Now there is no fulfilment of man's true vocation in

the order of nature. Man was made for more truth than he can see with his own unaided intelligence, and for more love than his will alone can achieve and for a higher moral activity than human prudence ever planned.

The prudence of the flesh is opposed to the will of God. The works of the flesh will bury us in hell. If we know and love and act only according to the flesh, that is to say, according to the impulses of our own nature, the things we do will rapidly corrupt and destroy our whole spiritual being.

In order to be what we are meant to be, we must know Christ, and love Him, and do what He did. Our destiny is in our own hands, since God has placed it there and given us His grace to do the impossible. It remains for us to take up courageously and without hesitation the work He has given us, which is the task of living our own life as Christ would live it in us.

It takes intrepid courage to live according to the truth, and there is something of martyrdom in every truly Christian life, if we take martyrdom in its original sense as a "testimony" to the truth, sealed in our own sufferings and in our blood.

4. Being and doing become one, in our life, when our life and being themselves are a "martyrdom" for the truth. In this way we identify ourselves with Christ, who said: "For this was I born, and for this came I into the world; that I should give testimony to the truth" (John 18 : 36). Our vocation is precisely this: to bear witness to the truth of Christ by laying down our lives at His bidding. Therefore, He added to the words we have just quoted: "Everyone that is of the truth, heareth my voice." And in another

place He said: "I know my sheep, and they know me" (John 10 : 14).

This testimony need not take the special form of a political and public death in defence of Christian truth or virtue. But we cannot avoid the "death" of our own will, of our own natural tendencies, of the inordinate passions of our flesh and of our whole selfish "being," in order to submit ourselves to what our own conscience tells us to be the truth and the will of God and the inspiration of the Spirit of Christ.

5. Therefore asceticism is unavoidable in Christian life. We cannot escape the obligation to deny ourselves. This obligation is made inevitable by the fact that the truth cannot live in us unless we freely and by our own volition recognize and cast out the falsity of sin from our own souls.

This is the one job that we alone can do, and we must have the courage to do it if we wish to live as we were meant to live, and find our true being in God. No one else can turn our minds to the truth, renounce error for us, convert our wills from selfishness to charity and from sin to God. The example and the prayers of others may help us to find our way in this work. But we alone can do it.

It is true that God is the One who produces in our hearts both our good desires and their effect, "for it is God who worketh in you both to will and to accomplish, according to His good will" (Philippians 2 : 13). Nevertheless, if we do not ourselves freely desire and manfully carry out His will, His grace will be without effect: since the effect of grace is to make us freely do His will.

Consequently, the truth of God lives in our souls more by the power of superior moral courage than by the light of an eminent intelligence. Indeed, spiritual intelligence itself depends on the fortitude and patience with which we sacrifice ourselves for the truth, as it is communicated to our lives concretely in the providential will of God.

6. The importance of courageous sacrifice, in accomplishing our work of finding and witnessing to the truth, cannot be over-emphasized. It is all-important. We cannot possess the truth fully until it has entered into the very substance of our life by good habits and by a certain perfection of moral activity. And we cannot act so without a terrible struggle against temptation, a struggle that divides our whole being against itself with conflicting loyalties. The greatest temptations are not those that solicit our consent to obvious sin, but those that offer us great evils masking as the greatest goods.

These apparent goods must be *sacrificed precisely as goods* before we can tell accurately whether they are good or evil. What is more, the things we are called upon to sacrifice may indeed remain perfectly good in themselves. That does not mean that our sacrifice of them is vain, or that we can take them back as soon as we have seen they are not evil. No: the fulfilment of every individual vocation demands not only the renouncement of what is evil in itself, but also *of all the precise goods that are not willed for us by God.*

It takes exceptional courage and integrity to make such a sacrifice. We cannot do it unless we are really seeking to do the will of God for His sake alone. The man who is content to keep from disobeying God, and to satisfy his own desires wherever there is nothing to prevent him from doing so, may indeed lead a life that is not evil: but his life will remain a sad confusion of truth and falsity and he will never have the spiritual vision to tell one clearly from the other. He will never fully live up to his vocation.

7. Our Father in Heaven has called us each one to the place in which He can best satisfy His infinite desire to do us good. His inscrutable choice of the office or the state of life or particular function to which we are called is not to be judged by the intrinsic merit of those offices and states but only by the hidden love of God. My vocation is the one I love, not because I think it is the best vocation in the Church, but because it is the one God has willed for me. If I had any evidence that He willed something else for me, I would turn to that on the instant. Meanwhile, my vocation is at once my will and His. I did not enter it blindly. He chose it for me when His inscrutable knowledge of my choice moved me to choose it for myself. I know this well enough when I reflect on the days when no choice could be made. I was unable to choose until His time had come. Since the choice has been made, there have been no signs in favour of changing it, and the presumption is that there will be no change. That does not mean there *cannot* be a change.

8. If we are called to the place in which God wills to do us the most good, it means we are called where we can best leave ourselves and find Him. The mercy of God demands to be known and recognized and set apart from everything else and praised and adored in joy. Every vocation is, therefore, at once a vocation to sacrifice and to joy. It is a call to the knowledge of God, to the recognition of God as our Father, to joy in the understanding of His mercy. Our individual vocation is our opportunity to

find that one place in which we can most perfectly receive the benefits of divine mercy, and know God's love for us, and reply to His love with our whole being.

That does not mean that our individual vocation selects for us a situation in which God will become visible to the eyes of our human nature and accessible to the feelings of our heart of flesh. On the contrary, if we are called where we will find Him we must go where flesh and blood will lose Him, for flesh and blood cannot possess the Kingdom of God (I Corinthians 15:50). God sometimes gives Himself to us where He seems to be taken away.

9. If I am called to the solitary life it does not necessarily mean that I will suffer more acutely in solitude than anywhere else: but that I will suffer more effectively. And for the rest, I will find there a greater joy because I shall know God in my sacrifice. In order to do this, I will not be too much aware of myself or of my sacrifice.

And there I will be most free to praise Him, even though my praise may be lowly and inarticulate and unworthy and poor. It will be most free, most mine, most Christ's. It will be the praise He seeks from me.

One who is not called to solitude will lose sight of God when he is alone and become troubled and turn upon himself in anxiety and in the end will become imprisoned in himself, unable to thank God or praise Him or do anything at all. He will have to look for Him somewhere else.

We know when we are following our vocation when our soul is set free from preoccupation with itself and is able to seek God and even to find Him, even though it may not appear to find Him. Gratitude and confidence and freedom from ourselves: these are signs that we have found our vocation and are living up to it even though everything else may seem to have gone wrong. They give us peace in any suffering. They teach us to laugh at despair. And we may have to.

10. There is something in the depths of our being that hungers for wholeness and finality. Because we are made for eternal life, we are made for an act that gathers up all the powers and capacities of our being and offers them simultaneously and forever to God. The blind spiritual instinct that tells us obscurely that our own lives have a particular importance and purpose, and which urges us to find out our vocation, seeks in so doing to bring us to a decision that will dedicate our lives irrevocably to their true purpose. The man who loses this sense of his own personal destiny and who renounces all hope of having any kind of vocation in life has either lost all hope of happiness or else has entered upon some mysterious vocation that God alone can understand.

Most human vocations tend to define their purpose not only by placing the one called in a definite relation to God, but also by giving him a set place among his fellow men. The vocation of each one of us is fixed just as much by the need others have for us as by our own need for other men and for God. Yet when I speak here of a need, I do not mean to exclude the untrammelled exercise of spiritual freedom. If I am called to the priesthood, it may be because the Church has need of priests and, therefore, that she has need of me. And it may also happen that my own peace and spiritual balance and the happiness of my whole life may ultimately depend on my becoming a priest. But the Church is not determined to accept me as a priest simply because she needs priests, nor am I forced to become a priest by the pressure of my own spiritual condition.

The freedom that is exercised in the choice of priestly vocations is a mystery hidden in God, a mystery that reaches out of the obscurity of God's Providence to select, sometimes, unlikely men to be "other Christs" and sometimes to reject those who are, in the eyes of men, best fitted for such a vocation.

11. What is the function of a priest in the world? To teach other men? To advise them? To console them? To pray for them? These things enter into his life, but they can be done by anyone. Every man in the world is called to teach and to advise and to console some other man, and we are all bound to pray for one another that we may be saved. These actions require no special priesthood, other than our baptismal participation in the priesthood of Christ, and they can be exercised even without this. Nor is the priest's distinctive vocation simply that he must be a man of God. The monk is a man of God and he does not have to be a priest.

The priest is called to be another Christ in a far more particular and intimate sense than the ordinary Christian or the monk. He must keep alive in the world the sacramental presence and action of the Risen Saviour. He is a visible human instrument of the Christ who reigns in Heaven, who teaches and sanctifies and governs the Church through His anointed priests. The words of the priest are not to be merely his own words or his own doctrine. They should always be the doctrine of the One who sent him. The action of the priest upon souls should come from something more than his own poor human power to advise and to console. Human though his acts may be, poor and deficient in themselves, they must be supported by the sacramental action of Jesus Christ and vivified by the hidden working of the Divine Spirit.

The priest is just as much sanctified by the actions he performs in the course of his sacred ministry as are those souls for whom he performs them. The Mass is, indeed, normally more fruitful for the priest who celebrates it than for any of those who assist at it. Indeed, one might say that the priest's holiness should be as great as the cumulative holiness of all those to whom he administers the sacraments. In any case, his vocation is to keep alive in the world the sanctity and the sanctifying power of the One High Priest, Jesus Christ.

This explains at once the beauty and the terror of the priestly vocation. A man, weak as other men, imperfect as they are, perhaps less well endowed than many of those to whom he is sent, perhaps even less inclined to be virtuous than some of them, finds himself caught, without possibility of escape, between the infinite mercy of Christ and the almost infinite dreadfulness of man's sin. He cannot help but feel in the depths of his heart something of Christ's compassion for sinners, something of the eternal Father's hatred of sin, something of the inexpressible love that drives the Spirit of God to consume sin in the fires of sacrifice. At the same time he may feel in himself all the conflicts of human weakness and irresolution and dread, the anguish of uncertainty and helplessness and fear, the inescapable lure of passion. All that he hates in himself becomes more hateful to him, by reason of his close union with Christ. But also by reason of his very vocation he is forced to face resolutely the reality of sin in himself and in others. He is bound by his vocation to fight this enemy. He cannot avoid the battle. And it is a battle that he alone can never win. He is forced to let Christ Himself fight the enemy in him. He must do battle on the ground chosen not by himself but by Christ. That ground is the hill of Calvary and the Cross. For, to speak plainly, the priest makes no sense at all in the world except to perpetuate in it the sacrifice of the Cross, and to die with Christ on the Cross for the love of those whom God would have him save.

12. Then there is the monastic vocation.

If the priest can be in some sense defined by other men's need of his sanctifying action in the world, this is less obviously true of the monk. For although it is true that the presence of every holy man in the world exercises a sanctifying effect, the monk does not exist precisely in order that others may be holy.

That is why it would be a mistake to assume that the essence of the monastic vocation is public prayer. The monk does, indeed, pray for other men and for the whole Church. But that is not the sole or even the main reason for his existence. Still less does the monk justify his existence by teaching, by writing, by the study of Scripture or of Gregorian chant, or by farming and raising cattle. There are plenty of cows in the world without monks to raise them.

It is true that the monastic vocation bears witness to the infinite transcendence of God, because it proclaims to the whole world that God has a right to call some men apart in order that they may live for Him alone. But in entering the monastery the monk should think of something more than this. Indeed, it would not be good for him to be too conscious of the fact that his sacrifice may still have some meaning to other men. If he dwells too long on the fact that the world remembers him, his very consciousness will re-establish the ties that he is supposed to have cut beyond recovery. For the essence of the monastic vocation is precisely this leaving of the world and all its desires and ambitions and concerns in order to live not only for God, but by Him and in Him, not for a few years but forever.

The one thing that most truly makes a monk what he is is this irrevocable break with the world and all that is in it, in order to seek God in solitude.

The world itself is even quicker to realize this fact than the monk who allows the purity of his vocation to be tarnished by concessions to the secular spirit. The first ones to condemn the monastery that has become infected with worldliness are those who, in the world, are themselves least monastic, for even those who have abandoned their religion often retain a high and exacting idea of religious perfection. St. Benedict saw that it was a matter of primary importance for the monk to "become a stranger to the ways of this world"—*a saeculi actibus se facere alienum*. But in establishing this principle, the Father of Western Monasticism was not simply thinking of public edification. He was thinking of the most urgent need of the monk's own soul.

13. The grace that calls a man to the monastery demands more than a physical change of environment. There is no genuine monastic vocation that does not imply, at the same time, a complete interior conversion. This conversion

can never be effected merely by a change of clothing or by the adoption of a stricter rule of life.

The habit does not make the monk, and neither do observances. The essential characteristic of a monastic vocation is that it draws the monk into solitude, to a life of self-renunciation and of prayer, in order to seek God alone. Where these features are not found, the vocation may, indeed, be a religious one, but it is not, properly speaking, monastic. It is a pity that monasteries of the ancient monastic Orders sometimes offer their subjects a life in which these elements are realized only in theory and not in fact.

Where the essentials of the monastic life are maintained, it matters little what accidental variations may be added to it. A monastic community can be physically and spiritually isolated from the world, can offer its monks a true life of prayer, and can at the same time maintain a school or some parishes without serious danger to the monastic spirit as it is interpreted by some branches of the Benedictine family. In the same way, the monastic life need not be disturbed by the presence of a small and well-ordered industry, carried on by the monks for their own support. But if the spirit of solitude and prayer and the exclusive love of God alone are not found in a monastery, it makes no difference how strict the rule may be, how inviolate the enclosure, how zealous the exterior zeal for liturgical functions: the men who live there are not really monks. The interior change, the *metanoia*, the turning to God which constitutes the very essence of the monastic calling, has not taken place in their souls.

14. The interior "conversion" that makes a monk will usually show itself outwardly in certain ways: obedience,

humility, silence, detachment, modesty. All of these can be summed up in the one word: *peace.*

The monastery is a house of God: therefore, a sanctuary of peace. True, this peace is bought at a price. It is not the tranquillity of a rich man's country home. It is the peace of poor men who are supernaturally content with their poverty, not because it delivers them from the worries and responsibilities of the world, nor yet because it helps them to lead a life that is essentially healthier and better balanced than the life of the world: but because it inexplicably puts them in possession of the God of all peace.

The peace of the monastic life is not to be accounted for by a natural and human explanation. Enter a monastery and see the life close at hand. You will find that what looks so perfect from the windows of the guest house is in reality full of the seams and cracks of human imperfection. The tempo of community life is not invariably serene. The order of the day can sometimes become unbalanced, surcharged, distracting as well as exhausting. Usages and observances are sometimes twisted into ridiculous formalities. There are moments when everything in the monastery seems to conspire to make peace and prayer impossible. These things inevitably ripple the surface of life in the best of communities. Their function is to remind us that the peace of the monks depends, ultimately, on something deep and hidden in their own souls. Monastic regularity is certainly most important in preserving peace. If regularity were to be lost forever, peace could not last long. But even where the life goes on according to rule, the rule alone is not sufficient to explain the peace of those who live by it. We must look deeper into the mystery of faith by which, in the secret recesses of their souls, the monks remain in possession of God no matter what may happen to disturb the surface of their lives.

The monastic life burns before the invisible God like a lamp before a tabernacle. The wick of the lamp is faith, the flame is charity, and the oil, by which the flame is fed, is self-sacrifice.

15. The monastic Orders are, of all religious Orders, the ones with the most ancient and the most monumental traditions. To be called to the monastic life is to be called to a way of sanctity that is rooted in the wisdom of the distant past, and yet is living and young, with something peculiarly new and original to say to the men of our own time. One cannot become a monk in the fullest sense of the word unless one's soul is attuned to the transforming and life-giving effect of the monastic tradition. And if this is true everywhere, it is especially true in America—a country in which men are not used to ancient traditions, and are not often ready to understand them.

What is the monastic tradition? It is the whole monastic way of life as it has been practised and handed down from generation to generation since the times of the first monks, the Fathers of the Egyptian desert, who in their turn felt that they were simply putting into practice the poverty and charity of the Apostles and first disciples of Christ. The monastic tradition is, therefore, a body of customs and attitudes and beliefs which sum up the whole wisdom of the monastic way of life. It tells the monk how to be a monk in the simplest and most effective way—the way in which monks have always been monks. But at the same time it tells him how to be a monk in the peculiar circumstances of his own time and place and culture.

Monastic tradition tells us, for example, what place prayer, reading, and work have in our lives. It shows us that hospitality is an important aspect of the monastic vocation. It teaches us that we must be men of penance and self-discipline, yet at the same time it teaches us the proper measure and discretion to be followed in all these things. It shows us clearly the relative unimportance of exterior observance when it is compared with the interior spirit and the real essentials of the monastic life. In a word, it sets everything in order in the monastic life.

Where the sense of monastic tradition is lacking, monks immediately begin to lead unbalanced lives. They are unable to learn true discretion. They cannot acquire a sense of proportion. They forget what they are supposed to be. They are not able to settle down and live at peace in the monastery. They cannot get along with their superiors or their brethren. Why do all these things happen? Because the monks who have never learned how to be real monks are driving themselves crazy trying to live the monastic life with the spirit and the methods appropriate to some other kind of life. Only a true sense of monastic tradition can preserve sanity and peace in monasteries. But this sense is not acquired automatically, especially in a monastery that has little or no sense of tradition. It must be learned. And it cannot be learned without direct contact with the channels of life through which it comes. That is why St. Benedict urged his own monks to read Cassian, St. Basil, and the Desert Fathers. But the reading of ancient monastic books is only one of these channels, and by no means the most important. The only way to become a monk is to live among real monks, and to learn the life from their example.

16. In this matter of monastic tradition, we must carefully distinguish between tradition and convention. In many monasteries there is very little living tradition, and yet the monks think themselves to be traditional. Why? Because they cling to an elaborate set of *conventions*. Convention and tradition may seem on the surface to be much the same thing. But this superficial resemblance only makes conventionalism all the more harmful. In actual fact, conventions are the death of real tradition as they are of all real life. They are parasites which attach themselves to the living organism of tradition and devour all its reality, turning it into a hollow formality.

Tradition is living and active, but convention is passive and dead. Tradition does not form us automatically: we have to work to understand it. Convention is accepted passively, as a matter of routine. Therefore convention easily becomes an evasion of reality. It offers us only pretended ways of solving the problems of living—a system of gestures and formalities. Tradition really teaches us to live and shows us how to take full responsibility for our own lives. Thus tradition is often flatly opposed to what is ordinary, to what is mere routine. But convention, which is a mere repetition of familiar routines, follows the line of least resistance. One goes through an act, without trying to understand the meaning of it all, merely because everyone else does the same. Tradition, which is always old, is at the same time ever new because it is always reviving—born again in each new generation, to be lived and applied in a new and particular way. Convention is simply the ossification of social customs. The activities of conventional people are merely excuses for *not* acting in a more integrally human way. Tradition nourishes the life of the spirit; convention merely disguises its interior decay.

Finally, tradition is creative. Always original, it always opens out new horizons for an old journey. Convention, on the other hand, is completely unoriginal. It is slavish imitation. It is closed in upon itself and leads to complete sterility.

Tradition teaches us how to love, because it develops and expands our powers, and shows us how to give ourselves to the world in which we live, in return for all that we have received from it. Convention breeds nothing but anxiety and fear. It cuts us off from the sources of all inspiration. It ruins our productivity. It locks us up within a prison of frustrated effort. It is, in the end, only the mask for futility and for despair. Nothing could be better than for a monk to live and grow up in his monastic tradition, and nothing could be more fatal than for him to spend his life tangled in a web of monastic conventions.

What has been said here of the monastic Orders applies even more strongly to some other forms of religious life in which tradition is less strong and convention can more easily hold sway.

17. We would be better able to understand the beauty of the religious vocation if we remembered that marriage too is a vocation. The religious life is a special way of sanctity, reserved for comparatively few. The ordinary way to holiness and to the fullness of Christian life is marriage. Most men and women will become saints in the married state. And yet so many Christians who are not called to religious life or to the priesthood say of themselves: "I have no vocation!" What a mistake! They have a wonderful vocation, all the more wonderful because of its relative

freedom and lack of formality. For the "society" which is the family lives beautifully by its own spontaneous inner laws. It has no need of codified rule and custom. Love is its rule, and all its customs are the living expression of deep and sincere affection. In a certain sense, the vocation to the married state is more desirable than any other, because of the fact that this spontaneity, this spirit of freedom and union in charity is so easily accessible, for the ordinary man, in family life. The formalism and artificiality which creep into religious communities are with difficulty admitted into the circle of a family where powerful human values triumphantly resist the incursions of falsity.

Married people, then, instead of lamenting their supposed "lack of vocation," should highly value the vocation they have actually received. They should thank God for the fact that this vocation, with all its responsibilities and hardships, is a safe and sure way to become holy without being warped or shrivelled up by pious conventionalism. The married man and the mother of a Christian family, if they are faithful to their obligations, will fulfil a mission that is as great as it is consoling: that of bringing into the world and forming young souls capable of happiness and love, souls capable of sanctification and transformation in Christ. Living in close union with God the Creator and source of life, they will understand better than others the mystery of His infinite fecundity, in which it is their privilege to share. Raising children in difficult social circumstances, they will enter perhaps more deeply into the mystery of divine Providence than others who, by their vow of poverty, ought ideally to be more directly dependent on God than they, but who in fact are never made to feel the anguish of insecurity.

18. All vocations are intended by God to manifest His love in the world. For each special calling gives a man some particular place in the Mystery of Christ, gives him something to do for the salvation of all mankind. The difference between the various vocations lies in the different ways in which each one enables men to discover God's love, appreciate it, respond to it, and share it with other men. Each vocation has for its aim the propagation of divine life in the world.

In marriage, God's love is made known and shared under the sacramentalized veils of human affection. The vocation to marriage is a vocation to a supernatural union which sanctifies and propagates human life and extends the Kingdom of God in the world by bringing forth children who will be members of the Mystical Christ. All that is most human and instinctive, all that is best in man's natural affections, is here consecrated to God and becomes a sign of divine love and an occasion of divine grace.

In married life, divine love is more fully incarnate than in the other vocations. For that reason it is easier to apprehend, easier to appreciate. But its extension, being less spiritual, is less wide. The sphere of action of the father and mother's love extends only to their own children and to their relatives and to a circle of friends and associates.

In order to extend the effectiveness of divine charity, the other vocations progressively spiritualize our human lives and actions in order to spread them over a wider and wider area. So, in the active religious life or in the secular priesthood the physical expression of human love is sacrificed, family life is given up, and the potentialities of love thus set free are extended to a whole parish or to a hospital or a school. In the active life instinctive human affections are consecrated to God more fully than in family life, and in a less incarnate fashion. But it is, nevertheless, still easy to see and appreciate the action of God's love in the corporal works of mercy—care of the sick and the poor, as well as in the tender care of homeless children, of the aged, and so on. Here too the labours and difficulties and sacrifices of the life bring with them a corresponding protection of human values in the soul of the one "called." In dealing with other people, one retains one's sense of relatedness and integration.

In the contemplative life the problems and difficulties are more interior and also much greater. Here divine love is less incarnate. We must apprehend it and respond to it in a still more spiritual way. Fidelity is much more difficult. The human affections do not receive much of their normal gratification in a life of silence and solitude. The almost total lack of self-expression, the frequent inability to "do things for" other people in a visible and tangible way, can sometimes be a torture and lead to great frustration. That is why the purely contemplative vocation is not for the immature. One has to be very strong and very solid to live in solitude.

Fortunately, the monastic life is not so purely contemplative that it does not provide for a certain amount of activity and self-expression. Living and working together in the monastic community, the monks normally preserve their sense of relatedness and do not lose their humanity. On the contrary, if they are faithful to the spirit of their rule, they will find human affection deepened and spiritualized into a profound union of charity which is no longer dependent on personal moods and fancies. And then they will come to realize something of their mission to embrace the whole world in a spiritual affection that is not limited in time or in space.

19. The higher one ascends in the scale of vocations the more careful the selection of the candidates must be. Normally, in the married life, selection takes care of itself: the will of God can be incarnate in a decision based on natural attraction. In the active life attraction and aptitude normally go together, and the one "called" can be accepted on the basis of his ability to do the required work in peace and with spiritual joy.

Normally, more than half the people who present themselves for admission to contemplative monasteries have no vocation. "Attraction" to the contemplative life is a much less serious criterion of vocation than attraction to the active life. The stricter and more solitary a contemplative Order may be, the greater will be the gap between "attraction" and "aptitude." That is especially true in a time like ours, in which men cannot find the normal amount of silence and solitude that human nature requires for its sound functioning. There are perhaps very many nuns and brothers in active Orders who have perfectly good active vocations, but who are so overworked and so starved for a normal life of prayer that they imagine they need to become Trappists or Carthusians. In some cases the solution may indeed be a transit to an enclosed Order, but more often all that is required is a proper adjustment in their own religious institute. The problem of such adjustments is too big even to be mentioned here.

The higher a man ascends in the scale of vocations the more he must be able to spiritualize and extend his affections. To live alone with God, he must really be able to live *alone*. You cannot live alone if you cannot stand loneliness. And you cannot stand loneliness if your desire for "solitude" is built on frustrated need for human affection. To put it in plain language, it is hopeless to try to live your life in a cloister if you are going to eat your heart out thinking that nobody loves you. You have to be able to disregard that whole issue, and simply love the whole world in God, embracing all your brethren in that same pure love, without seeking signs of affection from them and without caring whether or not you ever get any. If you think this is very easy, I assure you that you are mistaken.

20. To say that the enclosed, contemplative life is harder than the active life is not to say that the contemplative works harder, or that he has greater responsibilities and obligations to meet. The contemplative life is in many respects easier than the active life. But it is not easier to live it *well*.

It is relatively easy to "get by" in a contemplative monastery, to keep the rules, to be at the right place at the right time, and to go through all the motions. Admittedly the routine is laborious and tiresome, but you can get used to it. What is hard is not the business of putting forth physical effort, but the work of *really leading an interior life of prayer* underneath all the externals.

The mere fact that *everything* in a contemplative monastery is supposed to be geared for a life of prayer is precisely what makes it difficult, for those who do not have true vocations, to live fruitfully in the cloister. It is not too hard for them to lead lives of prayer when there is more working than praying in the daily round of duties. In a life where all is prayer, those who do not have a special contemplative vocation often end up by praying less than they would actually do in the active life.

21. Attraction to a certain kind of life and the ability to lead that life are not yet sufficient to establish that one has a vocation. Indeed, the element of attraction, which may be important in many cases, is not always in evidence. A man can be called to the priesthood and still have a sensible repugnance for some of the aspects of his vocation. A Trappist vocation does not necessarily exclude a shrinking from the austerity of Trappist life. The one thing that really decides a vocation is *the ability to make a firm decision* to embrace a certain state of life and *to act on that decision.*

If a person can never make up his mind, never firmly resolve to do what is demanded in order to follow a vocation, one can say that in all probability he has not received the vocation. The vocation may have been offered him: but that is something no one can decide with certainty. Whether or not he is resisting grace, the fact seems to be that he is "not called." But a calm and definite decision that is not deterred by obstacles and not broken by opposition is a good sign that God has given His grace to answer His call, and that he has corresponded to it.

In deciding a vocation one normally consults a spiritual director. His function is to give advice, encouragement, suggestions, and help. He may in certain cases forbid a person to carry out the idea of becoming a priest or a religious. But if he judges that a person can prudently follow a vocation, it remains for that person himself to make the final decision. No one, not even a director or confessor, not even an ecclesiastical superior, can decide for him. He must decide himself, since his own decision is the expression of his vocation. If he then applies for admission to a seminary or monastery, and if his application is accepted, he can say that he probably "has a vocation."

22. These thoughts on vocation are evidently incomplete. But there is one gap that needs to be filled in order to avoid confusion. We have spoken of the active and contemplative lives without, so far, referring to the vocation which St. Thomas rates higher than any other: the apostolic life in which the fruits of contemplation are shared with others.

Instead of speaking of this vocation in theory, let us rather look at its perfect embodiment in one of its greatest saints: Francis of Assisi. The stigmatization of St. Francis was a divine sign of the fact that he was, of all saints, the most Christlike. He had succeeded better than any other in the work of reproducing in his life the simplicity and the poverty and the love of God and of men which marked the life of Jesus. More than that, he was an Apostle who incarnated the whole spirit and message of the Gospels most perfectly. Merely to know St. Francis is to understand the Gospel, and to follow him in his true, integral spirit is to live the Gospel in all its fullness. The genius of his sanctity made him able to communicate to the world the teachings of Christ not in this or that aspect, not in fragments expanded by thought and analysis, but in all the wholeness of its existential simplicity. St. Francis was, as all saints must try to be, simply "another Christ."

His life did not merely reproduce this or that mystery of the life of Christ. He did not merely live the humble virtues of the divine infancy and of the hidden life at Nazareth. He was not merely tempted with Christ in the desert or weary with Him in the travels of His apostolate. He did not only work miracles like Jesus. He was not only crucified with Him. All these mysteries were united in the life of Francis, and we find them all in him, now singly and now together. The risen Christ lived again perfectly in this saint who was completely possessed and transformed by the Spirit of divine charity.

St. Thomas's phrase *contemplata aliis tradere* (to share with others the fruits of contemplation) is not properly understood unless we have in mind the image of a St. Francis walking the roads of medieval Italy, overflowing with the joy of a message that could only be communicated to him directly by the Spirit of God. The wisdom and the salvation preached by Francis were not only the overflow of the highest kind of contemplative life, but they were quite simply the expression of the fullness of the Christian Spirit—that is to say of the Holy Spirit of God.

No man can be an apostle of Christ unless he is filled with the Holy Ghost. And no man can be filled with the Holy Ghost unless he does what is normally expected of a man who follows Christ to the limit. He must leave all things, in order to recover them all in Him.

The remarkable thing about St. Francis is that in his sacrifice of everything he had also sacrificed all the "vocations" in a limited sense of the word. After having been edified for centuries by all the various branches of the Franciscan religious family, we are surprised to think that St. Francis started out on the roads of Umbria without the slightest idea that he had a "Franciscan vocation." And in fact he did not. He had thrown all vocations to the winds, together with his clothes and other possessions. He did not think of himself as an apostle, but as a tramp. He certainly did not look upon himself as a monk: if he had wanted to be a monk, he would have found plenty of monasteries to enter. He evidently did not go around conscious of the fact that he was a "contemplative." Nor was he worried by comparisons between the active and contemplative lives. Yet he led both at the same time, and with the highest perfection. No good work was alien to him—no work of mercy, whether corporal or spiritual, that did not have a place in his beautiful life! His freedom embraced everything.

Francis could have been ordained priest. He refused out of humility (for that too would have been a "vocation" and he was beyond vocations). Yet he had in fact the perfection and quintessence of the apostolic spirit of sacrifice and charity which are necessary in the life of every priest. It takes a moment of reflection to reconcile oneself to the thought that St. Francis never said Mass—a fact which is hardly believable to one who is penetrated with his spirit.

If there was any recognized vocation in his time that Francis might have associated with his own life, it was the vocation of hermit. The hermits were the only members of any set class of religious persons that he consistently imitated. He frequently went off into the mountains to pray and live alone. But he never thought that he had a "vocation" to do nothing else but that. He stayed alone as long as the Spirit held him in solitude, and then let himself be led back into the towns and villages by the same Spirit.

If he had thought about it, he might have recognized that his vocation was essentially "prophetic." He was like another Elias or Eliseus, taught by the Spirit in solitude, but brought by God to the cities of men with a message to tell them.

All the many facets of the vocation of a St. Francis show us that we are beyond the level of ordinary "states of life." But it is for that very reason that, whenever we speak of the "mixed life" or the "Apostolic vocation" we would do well to think of it in terms of a Francis or of an Elias. The "mixed life" is too easily reduced to its lowest common denominator, and at that level it is nothing more than a form of the active life. As such, it suffers by comparison with the contemplative life. Why? Because the dignity of the apostolic life, in the teaching of St. Thomas, flows not from the element of action that is in it but from the element of contemplation. A life of preaching without contemplation is nothing but an "active life," and though it may be very holy and meritorious, it cannot lay claim to the dignity ascribed by St. Thomas to the life which "shares with others the fruits of contemplation."

But in proportion as the mendicant friar approaches the ideal of his founder, in proportion as he *lives* the poverty and charity of Francis or Dominic, and plunges into the loving knowledge of God which is granted only to little ones, in proportion as he abandons himself to the Holy Spirit, he will far outstrip the contemplative perfection of those whose contemplation is given them for themselves alone.

9

The Measure of Charity

In the economy of divine charity we have only as much as we give. But we are called upon to give as much as we have, and more: as much as we are. So the measure of our love is theoretically without limit. The more we desire to give ourselves in charity, the more charity we will have to give. And the more we give the more truly we shall be. For the Lord endows us with a being proportionate to the giving for which we are destined.

Charity is the life and the riches of His Kingdom, and those are greatest in it who are least: that is, who have kept nothing for themselves, retaining nothing but their desire to give.

He who tries to retain what he is and what he has, and keep it for himself, buries his talent. When the Lord comes in judgment, this servant is found to have no more than he had at the beginning. But those who have made themselves less, by giving away what they had, shall be found both to be and to have more than they had. And to him who has most shall be given that which the unprofitable servant kept for himself.

"And he said to them that stood by: Take the pound away from him and give it to him that hath ten pounds. And they said to him: Lord, he hath ten pounds! But I say to you, that to every one that hath shall be given, and he shall abound: and from him that hath not, even that which he hath shall be taken away" (Luke 19: 24–26).

2. If I love my brother with a perfect love, I will want him to be free from every love but the love of God. Of all loves, charity alone is not possessive, because charity alone does not desire to be possessed. Charity seeks the greatest good of the one loved: and there is no greater good than charity. All other goods are contained in it. Charity is without fear: having given all that it has, it has nothing left to lose. It brings true peace, since it is in perfect concord with all that is good, and fears no evil.

Charity alone is perfectly free, always doing what it pleases: since it wills nothing except to love and cannot be prevented from loving. Without charity, knowledge is fruitless. Love alone can teach us to penetrate the hidden goodness of the things we know. Knowledge without love never enters into the inner secrets of being. Only love can truly know God as He is, for God is love.

Short of perfection, charity still feels fear: for it fears that it is not yet perfect. It is not yet perfectly free, since there is something left that it cannot do. It is still not at rest, for it is not yet perfectly given. It is still in the dark: for since it has not completely abandoned itself to God, it does not yet know Him. So it is still uncertain about finding Him in the things it knows.

No mere effort of ours can make our love perfect. The peace, certitude, liberty, fearlessness of pure love are gifts of God. Love that is not yet perfect must learn perfection by waiting upon His good pleasure, and bearing its own imperfection until the time is ripe for complete self-surrender. We cannot give unless there be someone to receive what we are giving. The gift of our charity is not perfect until God is ready to accept it. He makes us wait for the time of our whole-giving, so that by giving ourselves many times and in many ways in part, we may have more to surrender in the end.

3. We tend to identify ourselves with those we love. We try to enter into their own souls and become what they are, thinking as they think, feeling as they feel, and experiencing what they experience.

But there is no true intimacy between souls who do not know how to respect one another's solitude. I cannot be united in love with a person whose very personality my love tends to obscure, to absorb, and to destroy. Nor can I awaken true love in a person who is invited, by my love, to be drowned in the act of drowning me with love.

If we know God, our identification of ourselves with those we love will be patterned on our union with God, and subordinate to it. Thus our love will begin with the knowledge of its own limitations and rise to the awareness of its greatness. For in ourselves we will always remain separate and remote from one another, but in God we can be one with those we love.

We cannot find them in God without first perfectly finding ourselves in Him. Therefore we will take care not to lose ourselves in looking for them outside Him. For love is not found in the void that exists between our being and the being of the one we love. There is an illusion of unity between us when our thoughts, our words, or our emotions draw us out of ourselves and suspend us together for a moment over the void. But when this moment has ended, we must return into ourselves or fall into the void.

There is no true love except in God, who is the source both of our own being and of the being we love.

4. Charity is a love that fortifies the ones we love in the secrecy of their own being, their own integrity, their own contemplation of God, their own free charity for all who exist in Him.

Such love leads to God because it comes from Him. It leads to a union between souls that is as intimate as their own union with Him. The closer we are to God, the closer we are to those who are close to Him. We can come to understand others only by loving Him who understands them from within the depths of their own being. Otherwise we know them only by the surmises that are formed within the mirror of our own soul.

If we are angry, we will think them always angry. If we are afraid, we will think them alternately cowardly or cruel. If we are carnal, we will find our own carnality conveniently reflected in everyone who attracts us. And it is true that the instinct of connaturality may discover these things when the other person has not yet realized them to be there. So it is that we can attract others to us and draw the evil out of them by the force of our own passions. But in doing this we do not come to know them as they are: we only deform them so that we may know them as they are not. In doing so we bring an even greater deformity upon our own souls.

God knows us from within ourselves, not as objects, not as strangers, not as intimates, but as our own selves. His knowledge of us is the pure light of which our own self-knowledge is only a dim reflection. He knows us in Himself, not merely as images of something outside Him, but as "selves" in which His own self is expressed. He finds Himself more perfectly in us than we find ourselves.

He alone holds the secret of a charity by which we can love others not only as we love ourselves, but as He loves them. The beginning of this love is the will to let those we love be perfectly themselves, the resolution not to twist them to fit our own image. If in loving them we do not love what they are, but only their potential likeness to ourselves, then we do not love them: we only love the reflection of ourselves we find in them. Can this be charity?

5. Do not ask me to love my brother merely in the name of an abstraction—"society," the "human race," the "common good." Do not tell me that I ought to love him because we are both "social animals." These things are so much less than the good that is in us that they are not worthy to be invoked as motives of human love. You might as well ask me to love my mother because she speaks English.

We need abstractions, perhaps, in order to *understand* our relations with one another. But I may understand the principles of ethics and still hate other men. If I do not love other men, I will never discover the meaning of the "common good." Love is, itself, the common good.

There are plenty of men who will give up their interests for the sake of "society," but cannot stand any of the people they live with. As long as we regard other men as obstacles to our own happiness, we are the enemies of society and we have only a very small capacity for sharing in the common good.

6. We are obliged to love one another. We are not strictly bound to "like" one another. Love governs the will:

"liking" is a matter of sense and sensibility. Nevertheless, if we really love others it will not be too hard to like them also.

If we wait for some people to become agreeable or attractive before we begin to love them, we will never begin. If we are content to give them a cold impersonal "charity" that is merely a matter of obligation, we will not trouble to try to understand them or to sympathize with them at all. And in that case we will not really love them, because love implies an efficacious will not only to do good to others exteriorly but also to find some good in them to which we can respond.

7. Some people never reveal any of the good that is hidden in them until we give them some of the good, that is to say, some of the charity, that is in ourselves.

We are so much the children of God that by loving others we can make them good and lovable, in spite of themselves.

We are obliged to become perfect as our heavenly Father is perfect (Matthew 5 : 48). That means that we do not regard the evil in others, but give them something of our own good in order to bring out the good He has buried in them.

A Christian does not restrain his desire for revenge merely in order that he himself may be good, but in order that his enemy may be made good also. Charity knows its own happiness, and seeks to see it shared by everyone.

Charity, in order to be perfect, needs an equal. It cannot be content to love others as inferiors, but raises them to its own level. For unless it can share everything with the beloved, charity is not at rest. So it cannot find contentment merely in its own perfection. It demands the perfection of all.

8. There is a difference between loving men in God and loving God in men. The two loves are the same: they are charity, which has God for its object and which, by either act, attains directly to Him.

Yet there is a significant difference in emphasis, a difference of "focus" which gives these two acts a different character.

A life in which we love God in men is necessarily an active life. But the contemplative loves men in God.

When we love God in men, we seek to discover Him over and over in one individual after another. When we love men in God, we do not seek them. We find them without seeking them in Him whom we have found. The first kind of love is active and restless. It belongs more to time and to space than the other, which already participates in the changeless peace of eternity.

All charity grows in the same way: by increasing its intensity. Yet the love of God in men also extends itself in every direction to find new soil for its roots. The love of men in God grows only in depth: it plunges deeper and deeper into God, and by that very fact enlarges its capacity to love men.

When we love God in other men, our charity seeks to make His life grow in them. It surrounds that growth with anxious care. Our love develops more and more in our own soul while we watch its maturing in the souls of others.

But when we love men in God, we seek God and find Him with our whole being, and the growth of our love is simply the constant renewal of this supernatural encounter—an ever greater fullness of knowledge and of immersion in Him. The more we are plunged in Him, the better we can recognize Him wherever He is to be found: and the readier we are to see Him in other men.

To say that our contemplative charity finds our brothers in God, rather than seeing God in our brothers, is to say that it does not watch anxiously over His growth in their souls, and grow with them: but that it grows by itself in Him and, as a consequence, finds that others are also growing along with it.

If I love other men in God I can find them without turning away from Him. If I seek God in other men I find Him without turning away from them. In either case, when charity is fully mature, the brother whom I love is not too much of a distraction from the God in whom my love for him terminates.

9. Jesus did not come to seek God in men. He drew men to Himself by dying for them on the Cross, in order that He might be God in them. All charity comes to a focus in Christ, because charity is His life in us. He draws us to Himself, unites us to one another in His Holy Spirit, and raises us up with Himself to union with the Father.

Philosophy, which is abstract, speaks of "society" and the "common good." Theology, which is supremely concrete, speaks of the Mystical Body of Christ and of the Holy Ghost. It makes a great difference whether you look at life from the point of view of philosophy or from that of theology.

The common good never protests when it is violated. But the Holy Ghost speaks for Himself, argues, protests, urges, and insists. The common good does not move our wills. But "the charity of Christ is poured forth in our hearts by the Holy Ghost who is given to us" (Romans 5 : 5). The common good is too vague and too tame to put our passions to death within us: it can do nothing to defend itself against them! But the Holy Ghost promulgates in our hearts a law of love and self-denial which kills our selfishness and raises us up as new men in Christ: "For if by the Spirit you mortify the deeds of the flesh, you shall live!" (Romans 8 : 13).

The common good gives us no strength, teaches nothing either about life or about God. It passively waits for our homage and makes no murmur if it receive none. But "the Spirit also helpeth our infirmity. For we know not what we should pray for as we ought, but the Spirit Himself asketh for us" (Romans 8 : 26). And the Father strengthens us by His Spirit "with might unto the inward man, so that Christ may dwell by faith in our hearts, and that we may be rooted and founded in charity" (Ephesians 3 : 16).

The common good can offer us nothing except a kind of universal compromise in which the interests of countless human beings like ourselves will appear to be realized without too much conflict. The common good extends our horizons, no doubt, but only in order to give us a kind of Siberian landscape to contemplate: it is a vast, abstract steppe without any particular features, flat, low, mournful, under a cold grey sky. No wonder that men find the "common good" so uninteresting that they will build any selfish structure that serves to break its monotony!

10. The Holy Spirit not only widens our horizons, He lifts us into an entirely different world—a supernatural

order, where, as the "Spirit of Promise," He makes known to us the things that are hidden for us in God. "We have received not the Spirit of this world, but the Spirit that is of God: that we may know the things that are given us from God . . . as it is written: Eye hath not seen, nor ear heard, nor hath it entered into the heart of man, what things God hath prepared for them that love Him. But to us God hath revealed them by His Spirit. For the Spirit searcheth all things: yea, the deep things of God" (I Corinthians 2 : 9–12).

The things that are revealed to us by the Holy Spirit are the true common good: they are the infinite good which is God Himself, and the good of all His creatures in Him. Hence the Holy Spirit is not merely a leveller of individual interests, an arbiter, a judge who decrees some great universal compromise. God is the highest good not only of the collectivity but also and more particularly of each person in it. That is why in Scriptural language there is such constant use of the analogy of Father and Son. "Whosoever are led by the Spirit of God, they are the sons of God" (Romans 8 : 14). "Behold what manner of charity the Father hath bestowed upon us, that we should be called and should be the sons of God" (I John 3 : 1). Those who tend to stress the Christian collectivity so much that they make it a kind of totalitarian state of the spirit obscure the great truth of Christian personalism which is absolutely fundamental in our idea of the Mystical Body of Christ.

If the relation of our souls with God are the relations of sons to a father, that already brings out quite clearly that we are not mere units in a collectivity—employees in a factory, subjects in a state, soldiers in an army. We are sons, with rights of our own, rights that are the object of a most special care on the part of our Father. And the greatest of these rights is that which makes us His sons, and entitles us to a particular and special love of our own as sons, as individuals, as persons.

But our sonship before God is not a mere metaphor, or a legal fiction. It is a supernatural reality. This reality is the work of the Holy Ghost who not only confers upon us certain rights in the eyes of God, but even heightens and perfects our personality to the point of identifying us, each individually, with the only-begotten Son of God, Christ, the Incarnate Word. Consequently each Christian is not only a person in his own right, but his own personality is elevated by identification with the one Person who is the object of all the Father's love: the Word of God. Each one of us becomes completely himself when, in the Spirit of God, he is transformed in Christ. "But we all, beholding the glory of the Lord with open face, are transformed into the same image from glory to glory, as by the Spirit of the Lord" (II Corinthians 3 : 18).

11. If I say that the Holy Spirit is the "common good" of the Church, it is because He is also the "common good" of the Father and the Son. He is the bond between Them and He is given to us in order that we may love the Father in the Son and be loved by Him as He loves His own Son.

The Holy Spirit makes us other Christs by doing in us the work He does in the soul of Christ: He comes to us as the love of the Father and the Son for us. He awakens in us love for the Father in the Son, by drawing us to Jesus. "By this is the spirit of God known. Every spirit which confesseth that Jesus Christ is come in the flesh is of God" (I John 4 : 2).

The Holy Spirit has, therefore, for His chief function to draw us into the mystery of the incarnation and of our redemption by the Word made flesh. He not only makes us understand something of God's love as it is manifested to us in Christ, but He also makes us live by that love and experience its action in our hearts. When we do so, the Spirit lets us know that this life and action are the life and action of Christ in us. And so the charity that is poured forth in our hearts by the Holy Spirit brings us into an intimate, experiential communion with Christ. It is only by the Holy Spirit that we truly know and love Jesus and come through Him to the knowledge and love of the Father.

That is why St. Paul calls the Holy Spirit the "Spirit of Christ" and says: "You are not in the flesh but in the spirit, if so be that the Spirit of God dwell in you. Now if any man have not the Spirit of Christ, he is none of His" (Romans 8 : 9).

The Spirit of Christ, the Holy Spirit, is the life of the Mystical Body of Christ, His Church. Just as the soul is the vital principle on which the unity and action of a physical organism depend, so the Holy Spirit is the principle of life, unity, and action which draws the souls of men together to live as one in the "Whole Christ." But the Holy Spirit does not act independently of Jesus's own will. On the contrary: poured out upon Christ without measure, the Holy Spirit is given to each one of us "according to the measure of the giving of Christ" (Ephesians 4 : 7). Each one of us has in his heart the charity that Christ confers upon him, according to his merits, and the Spirit dwells in us in obedience to the will of Christ, the Head and Sanctifier of the Mystical Body.

It is by the Holy Spirit that we love those who are united to us in Christ. The more plentifully we have received of the Spirit of Christ, the more perfectly we are able to love them: and the more we love them the more we receive of the Spirit. It is clear, however, that since we love them by the Spirit who is given to us by Jesus, it is Jesus Himself who loves them in us.

The measure of charity is, therefore, in itself infinite, because charity is the gift of a Divine and Infinite Person. But the actual measure of charity in our souls is the "measure" that we have received from Christ. The question next to be answered, then, is: How much can we receive?

12. First of all, in order to receive anything at all of the Holy Spirit and of His love, we must first be baptized: that is to say that we must enter either sacramentally or by martyrdom or at least by a most perfect desire into the Mystery of the Passion and Resurrection of Christ. We must yield our souls to the action of His love, without which we cannot be elevated above our own natural level to participate in the things of God.

No mere ascetic technique, no symbolic and purely human religious rite, can bring us within the sacramental orbit of the love of Christ. He Himself must cast out the spirit of evil from our soul by the "finger of God" which is the Holy Spirit. Jesus Himself must baptize us with the Holy Spirit and with fire in order to make us "new creatures." St. John the Baptist said of the Saviour: "I indeed baptize you with water unto penance, but he that shall come after me is mightier than I, whose shoes I am not worthy to bear: He shall baptize you in the Holy Ghost and fire" (Matthew 3 : 11). And Jesus Himself made it clear that this baptism of the Holy Spirit was to be identified with the sacramental baptism given by His Church: for it is He who acts in the person of the baptizing priest, and works invisibly within the soul of the catechumen, washing it with His Holy Spirit at the same time as the priest performs the sacramental ablution—the outward sign of the interior grace conferred upon the soul by Christ Himself. "And Jesus said: Amen I say to thee, unless a man be born again of water and the Holy Ghost he cannot enter into the Kingdom of God" (John 3 : 5).

It is necessary that a man be redeemed and set apart from "the world" before he can receive the Holy Spirit, because "the world" in the New Testament is the collective name for all those subjected to the desire of temporal and carnal things as ends in themselves. The world in this sense is ruled by selfishness and illusion, and the "prince of this world" is also the "father of lies." But the Paraclete whom Jesus obtains for us from the Father and sends to dwell in our hearts forever "is the Spirit of Truth whom the world cannot receive, because it seeth Him not nor knowest Him: but you shall know Him because He shall abide with you and be in you" (John 14 : 17).

13. Once we have the Spirit dwelling in our hearts, the measure of the giving of Christ corresponds to our own desire. For in teaching us of the indwelling of His Spirit of charity, Jesus always reminds us to ask, in order that we may receive. The Holy Spirit is the most perfect gift of the Father to men, and yet He is the one gift which the Father

gives most easily. There are many lesser things that, if we ask for them, may still have to be refused us. But the Holy Spirit will never be refused. "If then you being evil know how to give good gifts to your children, how much more will your Father in heaven give the good Spirit to them that ask Him?" (Luke 11 : 13).

14. The first thing we must do when we recognize the presence of God's grace in our hearts is to desire more charity. The desire for love is itself a beginning of love, and from the moment we desire more we already have more: and our desire is itself the pledge of even more to come. This is because an efficacious desire to love God makes us turn away from everything that is opposed to His will.

It is by desiring to grow in love that we receive the Holy Spirit, and the thirst for more charity is the effect of this more abundant reception.

The desire for charity is more than a blind hunger of the soul (although in certain circumstances it is very blind and very much of a hunger). It is clear-sighted in the sense that the intelligence enlightened by the Holy Spirit turns to the Father and asks for an increase of love in the name of the Son. That is to say that the desire for charity in a mature Christian soul is a lucid, deep, peaceful, active, and supremely fruitful knowledge of the Holy Trinity.

This charity knows the Holy and Most Blessed Trinity of three Persons in one God, not by straining to keep in view three distinct "units" at once, which would be as difficult as it is false, but by seeking the Son, hidden in the Father, by the love of the Holy Spirit. There is only one love which draws us into one God. But this love is, as

God Himself is, triune. For charity unites us to the three Divine Persons from whom it comes—the Father is its inexhaustible source, the Son the hearth of its splendid brightness, and the Holy Spirit the power of its eternal unity.

All this Jesus has taught us in very concrete terms. "I go to the Father, and whatsoever you shall ask the Father in my name, that will I do, that the Father may be glorified in the Son. . . . If you love me, keep my commandments, and I will ask the Father and He will give you another Paraclete, that He may abide with you forever" (John 14: 13–16).

This last text gives us another way in which the measure of our love is increased: by obedience. Love does the will of the beloved. In obeying the Holy Spirit we receive a great increase of His charity in our hearts. For charity is the divine life which makes us sons of God. The more we obey the Spirit, the more we are moved and live as sons of God, and the greater our capacity for being enlightened and strengthened by His inspirations. "For whosoever are led by the Spirit of God, they are the sons of God" (Romans 8: 14). The Vulgate has "*aguntur*" for "led" in this sentence, suggesting that the soul that obeys the interior movement of the Holy Spirit is driven and impelled by charity to act as a son of God. The "charity of Christ presseth us," says St. Paul in another place (II Corinthians 5: 14). For the Holy Ghost speaks with a strong and almost unbearable insistence in the souls He would drive to increase their love.

This gives us the secret of the measure of charity. It is not merely our own desire but the desire of Christ in His Spirit that drives us to grow in love. Those who seldom or never feel in their hearts the desire for the love of God and other men, and who do not thirst for the pure waters of desire which are poured out in us by the strong, living God, are usually those who have drunk from other rivers or have dug for themselves broken cisterns.

It is not that the Holy Spirit does not wish to move them with His love: but they have no relish for the interior and spiritual movement of a pure and selfless charity. The world cannot receive the Spirit of God because it cannot know Him, and it cannot know Him because it does not know how to taste and see that the Lord is full of delights for the soul of the man who obeys Him.

If we have no taste for the things of God, we can at least desire to have that taste, and if we ask for it, it will be given us. But at the same time we must deny ourselves the taste for other things which kill the desire for God.

And so that brings us to another element that determines the measure of our love for God: self-denial. We receive as much of the charity of Christ as we are willing to deny ourselves of any other love. The one who has most in the realm of the spirit is the one who loves least in the order of the flesh. I say the one who loves least, not the one who eats least or drinks least or sleeps least: and a good married man may have more love for God than a second-rate monk. But the test, in any event, is detachment of the will and the desire to renounce oneself completely in order to obey God. In brief, then, the measure of our charity is the measure of our desire: that in turn is measured by God's own desires, and we let Him have His desires in us when we deny ourselves our own.

15. Even though the divine life that is given to us in Christ, by the Holy Spirit, is essentially the same life that we shall lead in Heaven, the possession of that life can never give us perfect rest on earth. A Christian is essentially an exile in this world in which he has no lasting city. The very presence of the Holy Spirit in his heart makes him discontent with worldly and material values. He cannot place his trust in the things of this life. His treasure is somewhere else, and where his treasure is, his heart is also.

16. We are saved by hope for that which we do not see and we wait for it with patience.

The Holy Spirit is the One who fills our heart with this hope and this patience. If we did not have Him speaking constantly to the depths of our conscience, we could not go on believing in what the world has always held to be mad. The trials that seem to defy our hope and ruin the very foundations of all patience are meant, by the Spirit of God, to make our hope more and more perfect, basing it entirely in God, removing every visible support that can be found in this world. For a hope that rests on temporal power or temporal happiness is not theological. It is merely human, and has no supernatural strength to give us.

But the hope of Christians is not merely focused on Heaven. Heaven itself is only the prelude to the final consummation revealed by Christ. The doctrine of the general resurrection teaches us that the glory of God's love is to be in a certain sense the common good of all things, not only of the souls of those who are saved in Christ but also of their bodies and of the whole material universe.

St. Paul tells us that the whole world and all the creatures in it, having fallen with man and having become, like man, subject to vanity and corruption, also unconsciously await re-establishment and fulfilment in the glory of the general resurrection. "For the expectation of the creature waiteth for the revelation of the sons of God. . . . Because the creature also itself shall be delivered from the servitude of corruption, into the liberty of the glory of the children of God" (Romans 8 : 19–21).

Now in the context of Pauline theology, this means only one thing: that the world and those in it redeemed by Christ must share in the Resurrection of Jesus from the dead. The Resurrection of Christ is, therefore, the heart of the Christian faith. Without it, the death of Jesus on the Cross is no more than the tragedy of an honest man—the death of a Jewish Socrates. Without the Resurrection, the teaching of Jesus is simply a collection of incoherent fragments with a vague moral reference: the Gospels lose most of their meaning.

The teaching and the miracles of Christ were not meant simply to draw the attention of men to a doctrine and a set of practices. They were meant to focus our attention upon God Himself revealed in the Person of Jesus Christ. Once again, theology is essentially concrete. Far from being a synthesis of abstract truths, our theology is centred in the Person of Jesus Himself, the Word of God, the Way, the Truth, and the Life. To understand this theology we must receive the Holy Spirit, who reminds us of all that Christ has said and done, and who introduces us into the abyss of the "deep things of God." The perfection of this theology is eternal life, which is "to know the one true God and Jesus Christ whom He has sent" (John 17 : 3).

Without the Resurrection, there is no sharing in the

divine life. The death of Jesus on the Cross expiated our sins, but it was only after He had risen that He breathed upon His disciples, giving them the Holy Ghost with the power to forgive sin, to baptize, to teach and preach to all nations and to renew His life-giving sacrifice.

If Christ is not risen from the dead then it is futile to say that He lives in His Church and in the souls of all Christians. For when we say that Christ lives in us, we do not mean that He is present in our minds as a model of perfection or as a noble memory or as a brilliant example: we mean that by His Spirit He Himself becomes the principle of new life and new actions which are truly and literally His life and His actions as well as our own. It is no metaphor for the Christian to say with St. Paul: "I live, now not I, but Christ liveth in me" (Galatians 2 : 20).

But in the mind of St. Paul the Resurrection of Christ demands our resurrection also, and the two are so inseparable that "if there be no resurrection of the dead, then Christ is not risen again" (I Corinthians 15 : 13). It is clear, then, that the general resurrection is so fundamental a doctrine in the Christian faith that no man who does not accept it can truly call himself a Christian. For St. Paul adds: "If Christ be not risen again, then is our preaching vain and your faith is also vain" (I Corinthians 15 : 17).

If our whole faith rests on the Resurrection of Jesus, if the Holy Ghost comes to us only from the Risen Christ, and if the whole of God's creation looks to the general resurrection in which it will share in the glory of the sons of God, then for a Christian the "common good" is really centred in the Resurrection of Christ from the dead. Anyone who wants to penetrate into the heart of Christianity and to draw forth from it the rivers of living water that give joy to the City of God (Psalm 45 : 5) must enter into this mystery. And the Mystery of the Resurrection is simply the completion of the Mystery of the Cross. We cannot enter into this mystery without the help of the Holy Spirit. But if we do, then "the Spirit of Him that raised up Jesus from the dead shall quicken also your mortal bodies, because of His Spirit that dwelleth in you" (Romans 8 : 11). "For this corruptible body must put on incorruption, and this mortal must put on immortality. And when this mortal hath put on immortality, then shall come to pass that saying that is written: 'Death is swallowed up in victory' " (I Corinthians 15 : 53–54).

17. In summary, then, the measure of our charity is theoretically infinite, because it depends on God's charity toward us, and this is infinite. In concrete reality, God has shown His charity for us in the Person of Christ. We live in Christ by His Spirit, and we at last become perfect in charity when we share perfectly in the mystery of the Resurrection in which Christ made us participate in His divine Sonship.

We will be perfect Christians when we have risen from the dead.

10

Sincerity

We make ourselves real by telling the truth. Man can hardly forget that he needs to know the truth, for the instinct to know is too strong in us to be destroyed. But he can forget how badly he also needs to tell the truth. We cannot know truth unless we ourselves are conformed to it.

We must be true inside, true to ourselves, before we can know a truth that is outside us. But we make ourselves true inside by manifesting the truth as we see it.

2. If men still admire sincerity today, they admire it, perhaps, not for the sake of the truth that it protects, but simply because it is an attractive quality for a person to have. They like to be sincere not because they love the truth, but because, if they are thought to be sincere, people will love them. And perhaps they carry this sincerity to the point of injustice—being too frank about others and themselves, using the truth to fight the truth, and turning it to an instrument of ridicule in order to make others less loved. The "truth" that makes another man seem cheap hides another truth that we should never forget, and which would make him remain always worthy of honour in our sight. To destroy truth with truth under the pretext of being sincere is a very insincere way of telling a lie.

3. We are too much like Pilate. We are always asking, "What is truth?" and then crucifying the truth that stands before our eyes.

But since we have asked the question, let us answer it.

If I ask, "What is truth?" I either expect an answer or I do not Pilate did not. Yet his belief that the question did not require an answer was itself his answer. He thought the question could not be answered. In other words, he thought it was true to say that the question, "What is truth?" had no satisfactory answer. If, in thinking that, he thought there was no truth, he clearly disproved his own proposition by his very thought of it. So, even in his denial, Pilate confessed his need for the truth. No man can avoid doing the same in one way or another, because our need for truth is inescapable.

What, then, is truth?

Truth, in things, is their reality. In our minds, it is the conformity of our knowledge with the things known. In our words, it is the conformity of our words to what we think. In our conduct, it is the conformity of our acts to what we are supposed to be.

4. It is curious that our whole world is consumed with the desire to know what things are, and actually does find out a tremendous amount about their physical constitution, and verifies its findings—and still does not know whether or not there is such a thing as truth!

Objective truth is a reality that is found both within and outside ourselves, to which our minds can be conformed. We must know this truth, and we must manifest it by our words and acts.

We are not required to manifest everything we know,

for there are some things we are obliged to keep hidden from men. But there are other things that we must make known, even though others may already know them.

We owe a definite homage to the reality around us, and we are obliged, at certain times, to say what things are and to give them their right names and to lay open our thought about them to the men we live with.

The fact that men are constantly talking shows that they need the truth, and that they depend on their mutual witness in order to get the truth formed and confirmed in their own minds.

But the fact that men spend so much time talking about nothing or telling each other the lies that they have heard from one another or wasting their time in scandal and detraction and calumny and scurrility and ridicule shows that our minds are deformed with a kind of contempt for reality. Instead of conforming ourselves to what is, we twist everything around, in our words and thoughts, to fit our own deformity.

The seat of this deformity is in the will. Although we still may speak the truth, we are more and more losing our desire to live according to the truth. Our wills are not true, because they refuse to accept the laws of our own being: they fail to work along the lines demanded by our own reality. Our wills are plunged in false values, and they have dragged our minds along with them, and our restless tongues bear constant witness to the disorganization inside our souls—"the tongue no man can tame, an unquiet evil, full of deadly poison. By it we bless God and the Father, and we curse men who are made in the likeness of God. . . . Doth a fountain send forth out of the same hole sweet and bitter water?" (James 3 : 8–11).

5. Truthfulness, sincerity, and fidelity are close kindred. Sincerity is fidelity to the truth. Fidelity is an effective truthfulness in our promises and resolutions. An inviolate truthfulness makes us faithful to ourselves and to God and to the reality around us: and, therefore, it makes us perfectly sincere.

Sincerity in the fullest sense must be more than a temperamental disposition to be frank. It is a simplicity of spirit which is preserved by the *will* to be true. It implies an obligation to manifest the truth and to defend it. And this in turn recognizes that we are free to respect the truth or not to respect it, and that the truth is to some extent at our own mercy. But this is a terrible responsibility, since in defiling the truth we defile our own souls.

Truth is the life of our intelligence. The mind does not fully live unless it thinks straight. And if the mind does not see what it is doing, how can the will make good use of its freedom? But since our freedom is, in fact, immersed in a supernatural order, and tends to a supernatural end that it cannot even know by natural means, the full life of the soul must be a light and strength which are infused into it supernaturally by God. This is the life of sanctifying grace, together with the infused virtues of faith, hope, charity, and all the rest.

Sincerity in the fullest sense is a divine gift, a clarity of spirit that comes only with grace. Unless we are made "new men," created according to God "in justice and the holiness of truth," we cannot avoid some of the lying and double-dealing which have become instinctive in our natures, corrupted, as St. Paul says, "according to the desire of error" (Ephesians 4 : 22).

One of the effects of original sin is an instinctive prejudice in favour of our own selfish desires. We see things as they are not, because we see them centred on ourselves. Fear, anxiety, greed, ambition, and our hopeless need for pleasure all distort the image of reality that is reflected in our minds. Grace does not completely correct this distortion all at once: but it gives us a means of recognizing and allowing for it. And it tells us what we must do to correct it. Sincerity must be bought at a price: the humility to recognize our innumerable errors, and fidelity in tirelessly setting them right.

The sincere man, therefore, is one who has the grace to know that he may be instinctively insincere, and that even his natural sincerity may become a camouflage for irresponsibility and moral cowardice: as if it were enough to recognize the truth and do nothing about it!

6. How is it that our comfortable society has lost its sense of the value of truthfulness? Life has become so easy that we think we can get along without telling the truth. A liar no longer needs to feel that his lies may involve him in starvation. If living were a little more precarious, and if a person who could not be trusted found it more difficult to get along with other men, we would not deceive ourselves and one another so carelessly.

But the whole world has learned to deride veracity or to ignore it. Half the civilized world makes a living by telling lies. Advertising, propaganda, and all the other forms of publicity that have taken the place of truth have taught men to take it for granted that they can tell other people whatever they like provided that it sounds plausible and evokes some kind of shallow emotional response.

Americans have always felt that they were protected against the advertising business by their own sophistication. If we only knew how naïve our sophistication really is! It protects us against nothing. We love the things we pretend to laugh at. We would rather buy a bad toothpaste that is well advertised than a good one that is not advertised at all. Most Americans wouldn't be seen dead in a car their neighbours had never heard of.

Sincerity becomes impossible in a world that is ruled by a falsity that it thinks it is clever enough to detect. Propaganda is constantly held up to contempt, but in contemning it we come to love it after all. In the end we will not be able to get along without it.

This duplicity is one of the great characteristics of a state of sin, in which a person is held captive by the love for what he knows he ought to hate.

7. Your idea of me is fabricated with materials you have borrowed from other people and from yourself. What you think of me depends on what you think of yourself. Perhaps you create your idea of me out of material that you would like to eliminate from your own idea of yourself. Perhaps your idea of me is a reflection of what other people think of you. Or perhaps what you think of me is simply what you think I think of you.

8. How difficult it is for us to be sincere with one another when we do not know either ourselves or one another! Sincerity is impossible without humility and supernatural love. I cannot be candid with other men unless I understand myself and unless I am prepared to do everything possible in order to understand them.

But my understanding of them is always clouded by the reflection of myself which I cannot help seeing in them.

It takes more courage than we imagine to be perfectly simple with other men. Our frankness is often spoiled by a hidden barbarity, born of fear.

False sincerity has much to say, because it is afraid. True candour can afford to be silent. It does not need to face an anticipated attack. Anything it may have to defend can be defended with perfect simplicity.

The arguments of religious men are so often insincere, and their insincerity is proportionate to their anger. Why do we get angry about what we believe? Because we do not really believe it. Or else what we pretend to be defending as the "truth" is really our own self-esteem. A man of sincerity is less interested in defending the truth than in stating it clearly, for he thinks that if the truth be clearly seen it can very well take care of itself.

9. Fear is perhaps the greatest enemy of candour. How many men fear to follow their conscience because they would rather conform to the opinion of other men than to the truth they know in their hearts! How can I be sincere if I am constantly changing my mind to conform with the shadow of what I think others expect of me? Others have no right to demand that I be anything else than what I ought to be in the sight of God. No greater thing could possibly be asked of a man than this! This one just expectation, which I am bound to fulfil, is precisely the one they usually do not expect me to fulfil. They want me to be what I am in their sight: that is, an extension of themselves. They do not realize that if I am fully myself, my life will become the completion and the fulfilment of

their own, but that if I merely live as their shadow, I will serve only to remind them of their own unfulfilment.

If I allow myself to degenerate into the being I am imagined to be by other men, God will have to say to me, "I know you not!"

10. The delicate sincerity of grace is never safe in a soul given to human violence. Passion always troubles the clear depths of sincerity, except when it is perfectly in order. And passion is almost never perfectly in order, even in the souls of the saints.

But the clean waters of a lake are not made dirty by the wind that ruffles their surface. Sincerity can suffer something of the violence of passion without too much harm, as long as the violence is suffered and not accepted.

Violence is fatal to sincerity when we yield it our consent, and it is completely fatal when we find peace in passion rather than in tranquillity and calm.

Spiritual violence is most dangerous when it is most spiritual—that is, when it is least felt in the emotions. It seizes the depths of the will without any surface upheaval and carries the whole soul into captivity without a struggle. The emotions may remain at peace, may even taste a delight of their own in this base rapture. But the deep peace of the soul is destroyed, because the image of truth has been shattered by rebellion. Such is the violence, for example, of unresisted pride.

There is only one kind of violence that captures the Kingdom of Heaven. It is the seeming violence of grace, which is really order and peace. It establishes peace in the soul's depth even in the midst of passion. It is called "violent" by reason of the energy with which it resists passion and sets order in the house of the soul. This violence is the voice and the power of God Himself, speaking in our soul. It is the authority of the God of peace, speaking within us, in the sanctuary, in His holy place.

The God of peace is never glorified by human violence.

11. The truth makes us saints, for Jesus prayed that we might be "sanctified in truth." But we also read that "knowledge puffeth up"—*scientia inflat.*

How is it that knowledge can make us proud?

There is no truth in pride. If our knowledge is true, then it ought to make us humble. If humble, holy.

As soon as truth is in the intellect, the mind is "sanctified" by it. But if the whole soul is to be sanctified, the will must be purified by this same truth which is in the intelligence. Even though our minds may see the truth, our wills remain free to "change the truth of God into a lie" (Romans 1 : 25).

There is a way of knowing the truth that makes us true to ourselves and God, and, therefore, makes us more real and holier. But there is another way of receiving the truth that makes us untrue, unholy. The difference between these two lies in the action of our will.

If my will acts as the servant of the truth, consecrating my whole soul to what the intelligence has seen, then I will be sanctified by the truth. I will be sincere. "My whole body will be lightsome" (Matthew 6 : 22).

But if my will takes possession of truth as its master, as if the truth were my servant, as if it belonged to me by right of conquest, then I will take it for granted that I can do with it whatever I please. This is the root of all falsity.

The saint must see the truth as something to serve, not

as something to own and manipulate according to his own good pleasure.

12. In the end, the problem of sincerity is a problem of love. A sincere man is not so much one who sees the truth and manifests it as he sees it, but one who loves the truth with a pure love. But truth is more than an abstraction. It lives and is embodied in men and things that are real. And the secret of sincerity is, therefore, not to be sought in a philosophical love for abstract truth but in a love for real people and real things—a love for God apprehended in the reality around us.

It is difficult to express in words how important this notion is. The whole problem of our time is not lack of knowledge but lack of love. If men only loved one another they would have no difficulty in trusting one another and in sharing the truth with one another. If we all had charity we would easily find God. "For charity is of God, and everyone that loveth is born of God, and knoweth God" (I John 4 : 7).

If men do not love, it is because they have learned in their earliest childhood that they themselves are not loved, and the duplicity and cynicism of our time belongs to a generation that has been conscious, since its cradle, that it was not wanted by its parents.

The Church understands human love far better and more profoundly than modern man, who thinks he knows all about it. The Church knows well that to frustrate the creative purpose of human generation is to confess a love that is insincere. It is insincere because it is less than human, even less than animal. Love that seeks only to enjoy and not to create is not even a shadow of love. It has no power.

The psychological impotence of our enraged generation must be traced to the overwhelming accusation of insincerity which every man and woman has to confront, in the depths of his own soul, when he seeks to love merely for his own pleasure. A love that fears to have children for any motive whatever is a love that fears love. It is divided against itself. It is a lie and contradiction. The very nature of love demands that its own creative fulfilment should be sought *in spite of every obstacle*. Love, even human love, is stronger than death. Therefore, it is even more obvious that true love is stronger than poverty or hunger or anguish. And yet the men of our time do not love with enough courage to risk even discomfort or inconvenience.

Is it surprising that the Church should completely disregard all the economic arguments of those who think money and comfort are more important than love? The life of the Church is itself the highest form of love, and in this highest love all lesser loves are protected and enshrined by a divine sanction. It is inevitable that in a day when men are emptying human love of all its force and content, the Church should remain its last defender. But it is surely ironical that even the physical pleasure of human love should be more effectively protected by the wise doctrine of the Church than by the sophisms of those whose only apparent end is pleasure. Here, too, the Church knows what she is talking about when she reminds us that man is made of body and soul, and that the body fulfils its functions properly only when it is completely subjected to the soul and when the soul is subjected to grace, that is, to divine love. In the doctrine of the Church, the virtue of temperance is meant not to crush or divert the human instinct for pleasure, but to make reasonable pleasure serve its own end: to bring man to perfection and happiness in union with God. And, thus, the Church is bound by her own inflexible logic to leave man all the ordinate fullness of the pleasures that are necessary for the well-being of the person and of the community. She never considers pleasure merely as a "necessary evil" which has to be tolerated. It is a good that can contribute to man's sanctification: but it is a good that fallen man finds it extremely difficult to put to proper use. Hence the stringency of her laws. But let us not forget the purpose of those laws, which is to guarantee not only the rights of God but also the rights of man himself: and even the legitimate rights of man's own physical body.

Once again do not accuse me of exaggeration in tracing the problem of sincerity to its roots in human love. The selfishness of an age that has devoted itself to the mere cult of pleasure has tainted the whole human race with an error that makes all our acts more or less lies against God. An age like ours cannot be sincere.

13. Our ability to be sincere with ourselves, with God, and with other men is really proportionate to our capacity for sincere love. And the sincerity of our love depends in large measure upon our capacity to believe ourselves loved. Most of the moral and mental and even religious complexities of our time go back to our desperate fear that we are not and can never be really loved by anyone.

When we consider that most men want to be loved as if they were gods, it is hardly surprising that they should despair of receiving the love they think they deserve. Even the biggest of fools must be dimly aware that he is not worthy of adoration, and no matter what he may believe about his right to be adored, he will not be long in finding out that he can never fool anyone enough to make her adore him. And yet our idea of ourselves is so fantastically unreal that we rebel against this lack of "love" as though we were the victims of an injustice. Our whole life is then constructed on a basis of duplicity. We assume that others are receiving the kind of appreciation we want for ourselves, and we proceed on the assumption that since we are not lovable as we are, we must become lovable under false pretences, as if we were something better than we are.

The real reason why so few men believe in God is that they have ceased to believe that even a God can love them. But their despair is, perhaps, more respectable than the insincerity of those who think they can trick God into loving them for something they are not. This kind of duplicity is, after all, fairly common among so-called "believers," who consciously cling to the hope that God Himself, placated by prayer, will support their egotism and their insincerity, and help them to achieve their own selfish ends. Their worship is of little value to themselves and does no honour to God. They not only consider Him a potential rival (and, therefore, place themselves on a basis of equality with Him), but they think He is base enough to make a deal with them, and this is a great blasphemy.

14. If we are to love sincerely and with simplicity, we must first of all overcome the fear of not being loved. And this cannot be done by forcing ourselves to believe in some illusion, saying that we are loved when we are not. We must somehow strip ourselves of our greatest illusions about ourselves, frankly recognize in how many ways we

are unlovable, descend into the depths of our being until we come to the basic reality that is in us, and learn to see that we are lovable after all, in spite of everything!

This is a difficult job. It can only really be done by a lifetime of genuine humility. But sooner or later we must distinguish between what we are not and what we are. We must accept the fact that we are not what we would like to be. We must cast off our false, exterior self like the cheap and showy garment that it is. We must find our real self, in all its elemental poverty but also in its very great and very simple dignity: created to be a child of God, and capable of loving with something of God's own sincerity and His unselfishness.

Both the poverty and the nobility of our inmost being consists in the fact that it is a *capacity* for love. It can be loved by God, and when it is loved by Him, it can respond to His love by imitation—it can turn to Him with gratitude and adoration and sorrow; it can turn to its neighbour with compassion and mercy and generosity.

The first step in this sincerity is the recognition that although we are worth little or nothing in ourselves, we are potentially worth very much, because we can hope to be loved by God. He does not love us because we are good, but we become good when and because He loves us. If we receive this love in all simplicity, the sincerity of our love for others will more or less take care of itself. Centred entirely upon the immense liberality that we experience in God's love for us, we will never fear that His love could fail us. Strong in the confidence that we are loved by Him, we will not worry too much about the uncertainty of being loved by other men. I do not mean that we will be indifferent to their love for us: since we wish them to love in us the God who loves them in us. But we will never have to be anxious about their love, which in any case we do not expect to see too clearly in this life.

15. The whole question of sincerity, then, is basically a question of love and fear. The man who is selfish, narrow, who loves little and fears much that he will not be loved, can never be deeply sincere, even though he may sometimes have a character that seems to be frank on the surface. In his depths he will always be involved in duplicity. He will deceive himself in his best and most serious intentions. Nothing he says or feels about love, whether human or divine, can safely be believed, until his love be purged at least of its basest and most unreasonable fears.

But the man who is not afraid to admit everything that he sees to be wrong with himself, and yet recognizes that he may be the object of God's love precisely because of his shortcomings, can begin to be sincere. His sincerity is based on confidence, not in his illusions about himself, but in the endless, unfailing mercy of God.

16. Sincerity is, perhaps, the most vitally important quality of true prayer. It is the only valid test of our faith, our hope, and our love of God. No matter how deep our meditations, nor how severe our penances, how grand our liturgy, how pure our chant, how noble our thoughts about the mysteries of God: they are all useless if we do not really mean what we say. What is the good of bringing down upon ourselves the curses uttered by the ancient prophets and taken up again by Christ Himself: "Hypocrites, well hath Isaias prophesied of you, saying: 'This

people honoureth me with their lips, but their heart is far from me' " (Matthew 15 : 7–8) (cf. Isaias 29 : 13).

Since the monk is a man of prayer and a man of God, his most important obligation is sincerity. Everywhere in St. Benedict's Rule we are reminded of this. Those who are not true monks "lie to God by their tonsure," he says (*Rule*, chapter I). The first thing to be sought in a candidate for the monastic life is sincerity in seeking God—*si vere Deum quaerit.* One of the instruments of good works, by which the monk becomes a saint, is to "utter truth from his heart and from his lips," and another is that he should not desire to be called a saint without being one, but become a saint in all truth. In order that the truth of his virtue might be more certain, he must desire to manifest it in deeds rather than words. But above all in his prayer, his thoughts must agree with what he sings (*mens concordet voci*). Like Jesus Himself, St. Benedict prefers that prayer should be short and pure, rather than that the monk should multiply empty words or meditations without meaning.

The most important thing in prayer is that we present ourselves as we are before God as He is. This cannot be done without a generous effort of recollection and self-searching. But if we are sincere, our prayer will never be fruitless. Our sincerity itself establishes an instant contact with the God of all truth.

II

Mercy

How close God is to us when we come to recognize and to accept our abjection and to cast our care entirely upon Him! Against all human expectation He sustains us when we need to be sustained, helping us to do what seemed impossible. We learn to know Him, now, not in the "presence" that is found in abstract consideration—a presence in which we dress Him in our own finery—but in the emptiness of a hope that may come close to despair. For perfect hope is achieved on the brink of despair when, instead of falling over the edge, we find ourselves walking on the air. Hope is always just about to turn into despair, but never does so, for at the moment of supreme crisis God's power is suddenly made perfect in our infirmity. So we learn to expect His mercy most calmly when all is most dangerous, to seek Him quietly in the face of peril, certain that He cannot fail us though we may be upbraided by the just and rejected by those who claim to hold the evidence of His love.

2. *Cum vero infirmor, tunc potens sum.* "When I am weak, then I am strong" (II Corinthians 12 : 10).

Our weakness has opened Heaven to us, because it has brought the mercy of God down to us and won us His love. Our unhappiness is the seed of all our joy. Even sin has played an unwilling part in saving sinners, for the infinite mercy of God cannot be prevented from drawing the greatest good out of the greatest evil. Sin was destroyed in the midst of the sin of those who thought they could destroy Christ. Sin can never do anything good. It cannot even destroy itself, which would indeed be a great good. But the love of Christ for us, and the mercy of God, destroyed sin by taking upon itself the burden of all our sins and by paying the price that was due to them. So the Church sings that Christ died on the tree of the Cross that life might arise from the same stem from which death had first grown: *ut unde mors oriebatur, inde vita resurgeret.*

The Christian concept of mercy is, therefore, the key to the transformation of a whole universe in which sin still seems to reign. For the Christian does not escape evil, nor is he dispensed from suffering, nor is he withdrawn from the influence and from the effects of sin: nor is he himself impeccable. He too, unfortunately, can sin. He has not been completely delivered from evil. Yet his vocation is to deliver the whole world from evil and to transform it in God: by prayer, by penance, by charity, and, above all, by mercy. God, who is all-holy, not only has had mercy on us, but He has given His mercy into the hands of potential sinners in order that they may be able to choose between good and evil, and may overcome evil with good, and may earn His mercy for their own souls by having mercy on others.

God has left sin in the world in order that there may be forgiveness: not only the secret forgiveness by which He Himself cleanses our souls, but the manifest forgiveness by which we have mercy on one another and so give expression to the fact that He is living, by His mercy, in our own hearts.

3. "Blessed are they that mourn." Can this be true? Is there any greater wretchedness than to taste the dregs of our own insufficiency and misery and hopelessness, and to know that we are certainly worth nothing at all? Yet it is blessed to be reduced to these depths if, in them, we can find God. Until we have reached the bottom of the abyss, there is still something for us to choose between all and nothing. There is still something in between. We can still evade the decision. When we are reduced to our last extreme, there is no further evasion. The choice is a terrible one. It is made in the heart of darkness, but with an intuition that is unbearable by its angelic clarity: when we who have been destroyed and seem to be in hell miraculously choose God!

4. Only the lost are saved. Only the sinner is justified. Only the dead can rise from the dead, and Jesus said, "I came to seek and to save that which was lost" (Luke 9 : 10).

5. Some men are only virtuous enough to forget that they are sinners without being wretched enough to remember how much they need the mercy of God.

It is possible that some who have led bad lives on earth may be higher in Heaven than those who appeared to be good in this life. What is the value of a virtuous life, if it be a life without love and without mercy? Love is the gift of God's mercy to human sorrow, not the reward of human self-sufficiency. Great sorrow for great evil is a tremendous gift of love, and takes us out of ourselves in ecstasy at the mercy of God: "Because she has loved much, much has been forgiven."

Yet the best thing of all is the love of the Virgin Mother

of God, who never offended Him, but who received from Him the greatest mercy of all: that of knowing her own nothingness in the midst of the greatest perfection, and of being the poorest of all the saints because she was the richest.

Her charity corresponded perfectly to her humility. She who was lowest in her own eyes saw without tremor that she was highest in God's eyes. She was glad of this because He was glad of it, and for no other reason.

Thus, she who was weakest became most powerful and overthrew the pride of all the strongest angels who had fallen because they wanted their power to be their own, as if it had come from themselves and belonged to no other. She, on the other hand, had no power of her own, but she had received the greatest mercy, had been the most loved and was, therefore, the most worthy of love and the most capable of loving God in return. By the power of what she did not have, she has received power over us all, and we belong to her by the right of God's mercy: in whose distribution she is sovereign, by her prayers.

6. In the holiness of God, all extremes meet—infinite mercy and justice, infinite love and endless hatred of sin, infinite power and limitless condescension to the weakness of His creatures. His holiness is the cumulation of all His other attributes, His being in its infinite transcendency, His otherness and utter difference from every other being.

Yet the supreme manifestation of God's holiness is the death of Christ on the Cross. Here, too, all extremes meet. And here man, who has run away from God and buried himself in corruption and in death in order not to see the holiness of His face, finds himself confronted, in death itself, with the Redeemer who is his life.

We must adore and acknowledge God's holiness by desiring Him to have mercy on us, and this is the beginning of all justice. To desire Him to be merciful to us is to acknowledge Him as God. To seek His pity when we deserve no pity is to ask Him to be just with a justice so holy that it knows no evil and shows mercy to everyone who does not fly from Him in despair.

Perfect hope in the mercy of God is the prerogative of those who have come closest to His sanctity. And because they are closest to holiness they seem to themselves to be farthest away from it: for the contrast and the opposition between holiness and themselves is felt to be unbearable. Under such circumstances it seems like presumption to hope at all, yet hope is an imperative necessity in their souls, because they are seized and possessed by the inexorable holiness of God. Therefore, while they see that it is seemingly impossible for their sins to be condoned, they are overwhelmed by the gratuitous actuality of their forgiveness. Such forgiveness could not come from anyone but God.

7. The power that manifests itself in our weakness is the power that was the strength of Christ's weakness—the love of the Father, who raised Him from the dead. Jesus went down into the dust of death in order that the power of His Resurrection might be manifest in our own lives. This power is seen, not in our natural gifts, not in talents or human wisdom or in man's strength. It is made evident only in the contest between what *appears* in us—what is human and our own—and by what does not appear: the secret power of grace.

8. "Blessed are the merciful, for they shall obtain mercy" (Matthew 5 : 7).

We can have the mercy of God whenever we want it, by being merciful to others: for it is God's mercy that acts on them, through us, when He leads us to treat them as He is treating us. His mercy sanctifies our own poverty by the compassion that we feel for their poverty, as if it were our own. And this is a created reflection of His own divine compassion, in our own souls. Therefore, it destroys our sins, in the very act by which we overlook and forgive the sins of other men.

Such compassion is not learned without suffering. It is not to be found in a complacent life, in which we platonically forgive the sins of others without any sense that we ourselves are involved in a world of sin. If we want to know God, we must learn to understand the weaknesses and sins and imperfections of other men as if they were our own. We must feel their poverty as Christ experienced our own.

9. We can only get to Heaven by dying for other people on the cross. And one does not die on a cross by his own unaided efforts. He needs the help of an executioner. We have to die, as Christ died, for those whose sins are to us more bitter than death—most bitter because they are just like our own. We have to die for those whose sins kill us, and who are killed, in spite of our good intentions, by many sins of our own.

If my compassion is true, if it be a deep compassion of the heart and not a legal affair, or a mercy learned from a book and practised on others like a pious exercise, then my compassion for others is God's mercy for me. My patience with them is His patience with me. My love for them is His love for me.

10. When the Lord hears my prayer for mercy (a prayer which is itself inspired by the action of His mercy), then He makes His mercy present and visible in me by moving me to have mercy on others as He has had mercy on me. This is the way in which God's mercy fulfils His divine justice: mercy and justice seem to us to differ, but in the works of God they are both expressions of His love.

His justice is the love that gives to each one of His creatures the gifts that His mercy has previously decreed. And His mercy is His love, doing justice to its own exigencies, and renewing the gift which we had failed to accept.

11. The mercy of God does not suspend the laws of cause and effect. When God forgives me a sin, He destroys the guilt of sin, but the effects and the punishment of sin remain. Yet it is precisely in punishing sin that God's mercy most evidently identifies itself with His justice. Every sin is a violation of the love of God, and the justice of God makes it impossible for this violation to be perfectly repaired by anything but love. Now love is itself the greatest gift of God to men. Charity is our own highest perfection and the source of all our joy. This charity is the free gift of His mercy. Filling us with divine charity and calling us to love Him as He has first loved us and to love other men as He has loved us all, God's mercy makes it possible for us to give full satisfaction to His justice. The justice of God can, therefore, best be satisfied by the effects of His own mercy.

Those who refuse His mercy satisfy His justice in another way. Without His mercy, they cannot love Him. Without love for Him they cannot be "justified" or "made just." That is to say: they cannot conform to Him who is love. Those who have not received His mercy are in a state of injustice with regard to Him. It is their own injustice that is condemned by His justice. And in what does their injustice consist? In the refusal of His mercy. We come, then, in the end, to this basic paradox: that we owe it to God to receive from Him the mercy that is offered to us in Christ, and that to refuse this mercy is the summation of our "injustice." Clearly, then, only the mercy of God can make us just, in this supernatural sense, since the primary demand of God's justice upon us is that we receive His mercy.

12. Do you want to know God? Then learn to understand the weaknesses and imperfections of other men. But how can you understand the weaknesses of others unless you understand your own? And how can you see the meaning of your own limitations until you have received mercy from God, by which you know yourself and Him? It is not sufficient to forgive others: we must forgive them with humility and compassion. If we forgive them without humility, our forgiveness is a mockery: it presupposes that we are better than they. Jesus descended into the abyss of our degradation in order to forgive us after He had, in a sense, become lower than us all. It is not for us to forgive others from lofty thrones, as if we were gods looking down on them from Heaven. We must forgive them in the flames of their own hell, for Christ, by means of our forgiveness, once again descends to extinguish the aveng-

ing flame. He cannot do this if we do not forgive others with His own compassion. Christ cannot love without feeling and without heart. His love is human as well as divine, and our charity will be a caricature of His love if it pretends to be divine only, and does not consent to be human.

When we love others with His love, we no longer know good and evil (which was what the serpent promised) but only good. We overcome the evil in the world by the charity and compassion of God, and in so doing we drive all evil out of our own hearts. The evil that is in us is more than moral. There is a psychological evil, the distortion caused by selfishness and sin. Good moral intentions are enough to correct what is formally bad in our moral acts. But in order that our charity may heal the wounds of sin in our whole soul it must reach down into the farthest depths of our humanity, cleaning out all the infection of anxiety and false guilt that spring from pride and fear, releasing the good that has been held back by suspicion and prejudice and self-conceit. Everything in our nature must find its right place in the life of charity, so that the whole man may be lifted up to God, that the entire person may be sanctified and not only the intentions of his will.

13. God who is infinitely rich became man in order to experience the poverty and misery of fallen man, not because He needed this experience but because we needed His example. Now that we have seen His love, let us love one another as He has loved us. Thus His love will work in our hearts and transform us into Himself.

12

Recollection

RECOLLECTION is a change of spiritual focus and an attuning of our whole soul to what is beyond and above ourselves. It is a "conversion" or a "turning" of our being to spiritual things and to God. And because spiritual things are simple, recollection is also at the same time a simplification of our state of mind and of our spiritual activity. This simplification gives us the kind of peace and vision of which Jesus speaks when He says: "If thine eye be single thy whole body will be lightsome" (Matthew 6 : 22). Since this text refers principally to purity of intention, it reminds us that recollection does this also: it purifies our intention. It gathers up all the love of our soul, raises it above created and temporal things, and directs it all to God in Himself and in His will.

2. True recollection is known by its effects: peace, interior silence, tranquillity of heart. The spirit that is recollected is quiet and detached, at least in its depths. It is undisturbed because the passions are momentarily at rest. At most, they are allowed to trouble only the surface of the recollected soul. But since the fruits of recollection are produced by humility and charity and by most of the other fundamental Christian virtues, it is clear that true recollection cannot exist except when these virtues concur to give it substance and actuality.

3. Concentration is not recollection. The two can exist together, but ordinarily recollection means so much more than the focusing of thought upon a single clear point that it tends to diffuse thought by simplification, thus raising it above the level of tension and self-direction.

4. Recollection is more than a mere turning inward upon ourselves, and it does not necessarily mean the denial or exclusion of exterior things. Sometimes we are more recollected, quieter, simple and pure, when we see *through* exterior things and see God in them than when we turn away from them to shut them out of our minds. Recollection does not deny sensible things, it sets them in order. Either they are significant to it, and it sees their significance, or else they have no special meaning, and their meaninglessness remains innocent and neutral. For recollection brings the soul into contact with God, and His invisible presence is a light which at once gives peace to the eye that sees by it, and makes it see all things in peace.

5. Recollection should be seen not as an absence, but as a presence. It makes us, first of all, present to ourselves. It makes us present to whatever reality is most significant in the moment of time in which we are living. And it makes us present to God, present to ourselves in Him, present to everything else in Him. Above all, it brings His presence to us. And this is what gives meaning to all the other "presences" of which we have spoken.

6. First of all, we must be present to ourselves.

The cares and preoccupations of life draw us away from ourselves. As long as we give ourselves to these things, our minds are not at home. They are drawn out of their own reality into the illusion to which they tend. They let go of the actuality which they have and which they are, in order to follow a flock of possibilities. But possibilities have wings, and our minds must take flight from themselves in order to follow them into the sky. If we live with possibilities we are exiles from the present which is given us by God to be our own, homeless and displaced in a future or a past which are not ours because they are always beyond our reach. The present is our right place, and we can lay hands on whatever it offers us. Recollection is the only thing that can give us the power to do so. But before we explain that, let us return to the idea that recollection makes us present to ourselves.

As long as we are in this life, we both are and are not. We are constantly changing, and yet the person who changes is always the same person. Even his changes express his personality, and develop it, and confirm him for what he is.

A man is a free being who is always changing into himself. This changing is never merely indifferent. We are always getting either better or worse. Our development is measured by our acts of free choice, and we make ourselves according to the pattern of our desires.

If our desires reach out for the things that we were created to have and to make and to become, then we will develop into what we were truly meant to be.

But if our desires reach out for things that have no meaning for the growth of our spirit, if they lose themselves in dreams or passions or illusions, we will be false to ourselves and in the end our lives will proclaim that we have lied to ourselves and to other men and to God. We will judge ourselves as aliens and exiles from ourselves and from God.

In hell, there is no recollection. The damned are exiled not only from God and from other men, but even from themselves.

7. Recollection makes me present to myself by bringing together two aspects, or activities, of my being as if they were two lenses in a telescope. One lens is the basic substance of my spiritual being: the inward soul, the deep will, the spiritual intelligence. The other is my outward soul, the practical intelligence, the will engaged in the activities of life.*

When my practical and outward self is submissive and ordered to the deepest needs implanted in my inward being by nature and by grace, then my whole soul is in harmony with itself, with the realities around it, and with God. It is able to see things as they are, and it enables me to be aware of God. This makes me "present" to myself. That is to say that my outward self is alive to its true function as a servant of the spirit and of grace. It is aware of the mastery of grace, aware of the inward self, aware of its submission to the spirit and to grace, and aware of its power to work among outward things, and by this work to help the inward transformation of the spirit by grace.

But when the outward self knows only itself, then it is absent from my true self. It does not know its own inward spirit. It never acts according to the need and measure of my own true personality, which exists where my spirit is

* I do not mean that there are two souls in man with two sets of faculties. Man's one soul acts in two different ways, when dealing with the outward works and inward contemplation. Cf. St. Augustine, *De Trinitate, Lib*. XII.

wedded with the silent presence of the Lord's Spirit, and where my deep will responds to His gravitation toward the secrecy of the Godhead.

When I am not present to myself, then I am only aware of that half of me, that mode of my being which turns outward to created things. And then it is possible for me to lose myself among them. Then I no longer feel the deep secret pull of the gravitation of love which draws my inward self toward God. My will and my intelligence lose their command of the other faculties. My senses, my imagination, my emotions, scatter to pursue their various quarries all over the face of the earth. Recollection brings them home. It brings the outward self into line with the inward spirit, and makes my whole being answer the deep pull of love that reaches down into the mystery of God.

8. Recollection, then, makes me present to whatever is significantly real at each moment of my existence. The depths of my soul should always be recollected in God. When they are so, they do not necessarily prevent me from engaging in practical, outward activity. True, there are certain modes of union with God which interfere with exterior action. But recollection as such is compatible with physical and mental activity, and with any ordinate kind of work.

In order to be recollected in action I must not lose myself in action. And in order to keep acting, I must not lose myself in recollection. Hence recollected activity, in carrying out the will of God in the duties of my state of life, means a balance between interior purity and exterior attention. Both these are required. It is a shocking thing to see a monk do his work carelessly because he is trying to work and pray at the same time in such a way that he neither works nor prays properly.

We fail to balance interior purity and exterior attention when, in one way or another, we seek ourselves instead of God. If we seek only our own interests in our work, we will not be able to keep our hearts pure and recollected in an atmosphere of prayer. If we seek ourselves in prayer, we will not give the proper attention to our work.

The secret of recollected action is, first of all, detachment from ourselves and from the results either of our action or of our prayer. We must be detached from the results of our work, in order to deliver ourselves from the anxiety that makes us plunge into action without restraint. We must also be detached from the desire to see ourselves always recollected in God and to feel His presence in our hearts. That is to say, we must go to work for God, trusting that if we seek only to do His will, He will take care of our interior recollection, and make up for the distractions and failings that may creep into our activity.

If we begin our work with peace and recollection and with our hearts directed to God by prayer and a pure intention, we will avoid many baneful concerns and useless preoccupations as we proceed with our work. Scattered powers are easily fatigued. Spiritual fatigue proceeds from wasted and ill-applied effort. The disgust it generates slackens the powers of our soul, so that we waste even more effort and end in even greater fatigue.

9. Thought that does not proceed from recollection tends by its very nature to disperse our powers of thought and will. It may seem to seek recollection, but it cannot be helped by what it can never find. And if it is not strength-

ened by interior recollection, such thought must seek its strength elsewhere—in the vain power of excitement and interior tension. This tension is a form of unconscious idolatry, centred upon the illusion of a fixed idea, in which we place our trust. We impose this idea upon ourselves by violence, concentrate our thoughts upon it, and hug it to our hearts. This idea may be true in itself, but the violence of our own will gives it a disproportionate place in our own interior life. So it becomes an idol, drawing to itself the worship and attention and trust that we owe to God alone.

10. Anxiety is fatal to recollection because recollection depends ultimately on faith, and anxiety eats into the heart of faith. Anxiety usually comes from strain, and strain is caused by too complete a dependence on ourselves, on our own devices, our own plans, our own idea of what we are able to do. If we rely exclusively on our own efforts to keep ourselves recollected at work, our recollection will be forced and artificial. It will inevitably come in conflict with our work itself. When the Lord gives us work to do for Him, He does not demand that we work like angels, with our minds at the same time totally engrossed in Him and efficiently given to the work commanded. It is much simpler to use a little common sense, to set about working for the love of God, to keep ourselves detached from the anxieties and cares aroused by the work itself, and not to worry too much about our spiritual state while we are working. It is enough to think of one thing at a time, and the spiritual life is never helped by constant and exaggerated reflection upon ourselves.

At the same time it is not hard to work in a recollected

way if our intention is constantly purified by faith in the God for whom we are working, and by trust in His love, which keeps our souls united to Him even though we may not always be conscious of the union. We must remember that our experience of union with God, our feeling of His presence, is altogether accidental and secondary. It is only a side effect of His actual presence in our souls, and gives no sure indication of that presence in any case. For God Himself is above all apprehensions and ideas and sensations, however spiritual, that can ever be experienced by the spirit of man in this life.

11. Recollection also makes us present to God, and to ourselves in Him. The desire to preserve the deepest movements of our soul for God alone, to direct them away from ourselves and from His creatures, and concentrate them entirely on the fulfilment of His will, makes us in a special way present to God. True, we are always present to Him who sees all and keeps all things in existence by the very act that knows their existence. But we are more present to Him when we are aware of His nearness to us than when we ignore it. For then the presence is conscious and mutual: it is the presence of a person to a person. And it is only when we are thus present to Him that we truly discover ourselves as we really are. For when we are in the presence of God, seeing Him in His own light, which comes to us in the obscurity of faith, we also see, in this same light, that we are far different from what we thought ourselves to be in the light of our own ambition and self-complacency.

Here recollection becomes tinged with compunction and what the Fathers called "holy fear." Fear is the knowledge of ourselves in the presence of God's holiness. It is the knowledge of ourselves in His love, and it sees how far we are from being what His love would have us be. It knows who He is and who we are!

But fear that is holy cannot fear love. It fears the discrepancy between itself and love, and flies to hide itself in the abyss of light which is God's love and His perfection.

This fear is sometimes absolutely necessary to keep our recollection from becoming perverted with a false sweetness and from the presumptuous self-assurance which takes the grace of God for granted, which ceases to fear delusion, and which becomes complacent in the thought of our own virtue and our own lofty degree of prayer. Such complacency falls imperceptibly, like an impenetrable curtain, between ourselves and God. He departs from us, leaving us with a terrible illusion.

The more intense, strained, and prolonged our recollection, the greater the danger that we may fall into this illusion. It becomes easy enough, after a time, to recollect ourselves without actually entering into any real contact with God at all. Such recollection is nothing but a psychological trick. It is an act of introversion which can be learned with a little effort. It opens the door to a dark, silent, comfortable interior room in which nothing ever happens and in which there are no more troubles because we have managed to find the switch that turns off all our intellectual activity at its source. This is not prayer, and although it may be restful and even beneficial to us for a little while, if we become attached to it and carry it too far it can lead to very serious harm.

Recollection without faith confines the spirit in a prison without light or air. Interior asceticism should not end by locking us in such a prison. It would only defeat all the purposes of God's grace by doing so. Faith establishes us in recollection not by setting limits to the activity of the soul, but by removing all the limitations of our natural intelligence and will, freeing the mind from doubt and the will from hesitation, so that the spirit is let loose by God and plunges into the depths of His invisible freedom.

12. Recollection is almost the same thing as interior solitude. It is in recollection that we discover the finite solitude of our own heart and the infinite solitude of God dwelling within us. Unless these vast horizons have opened out in the centre of our lives, we can hardly see things in perspective. Our judgments are not in proportion with things as they are. But the spiritual man, says St. Paul, judges all things. He does so because he is isolated from them by his detachment, by his poverty, by his humility, by his nothingness. Therefore, he sees them only in God. To see them thus is to judge them as God Himself judges them.

Recollection brings us, then, to an interior solitude which is something more than either the desire or the fact of being alone. We become solitaries not when we realize how alone we are, but when we sense something of the solitude of God. His solitude isolates us from everything around us, and yet makes us all the more truly the brothers of all things.

We cannot live for others until we have entered this solitude. If we try to live for them without first living entirely for God, we risk plunging with them all into the abyss.

13. How many there are who have solitude and do not love it, because their solitude is without recollection! It

is only loneliness. It does nothing to bring them to themselves. They are alone because in their solitude they are separated from God, and from other men, and even from themselves. They are like souls wandering out of hell and finding their way by mistake into Heaven, only to discover that Heaven is more of a hell to them than hell itself. So it is with those who are forced into the heaven of solitude and cannot taste its joy because they know no recollection.

The man who fears to be alone will never be anything but lonely, no matter how much he may surround himself with people. But the man who learns, in solitude and recollection, to be at peace with his own loneliness, and to prefer its reality to the illusion of merely natural companionship, comes to know the invisible companionship of God. Such a one is alone with God in all places, and he alone truly enjoys the companionship of other men, because he loves them in God in whom their presence is not tiresome, and because of whom his own love for them can never know satiety.

14. False recollection occurs when we try by our own efforts to block out all material things, to isolate ourselves from people and nature by main force, hoping that there will be nothing left in our soul but God. When we attempt this, we usually divide our being against itself, call one half (the one we like) God, and call the other our "nature" or our "self." What madness, what a waste of effort, to try to rest in one half of our being, calling it "God," and lock the other out of doors! Our being resists this division, and engages in what we think is a war between light and darkness. But this struggle is only the battle of an illusion

against an illusion. Such battles are too often waged in monasteries, where God calls men not to embrace illusion but to abandon it, that they may discover what is real.

False recollection is inevitable without humility. For humility teaches us to accept ourselves as we are, restrains our pride from forcing ourselves to be what we are not. It is better to be content with what is low and unassuming in the spiritual life, for in this life all voluntary poverty is a spiritual enrichment. To be content with a low degree of prayer is to be enriched in prayer. Such contentment is better than the pride that insists on striving for the intellectual purity of the angels before it has even learned to be maturely human.

13

"My Soul Remembered God"

THE "remembering" of God, of which we sing in the Psalms, is simply the rediscovery, in deep compunction of heart, that God remembers us. In a sense, God cannot be remembered. He can only be discovered.*

We know Him because He knows us. We know Him when we discover that He knows us. Our knowledge of Him is the effect of His knowledge of us. The experience is always one of new wonder that He is mindful of us. "What is man that Thou art mindful of him? Or the son of man that Thou visitest him?" (Psalm 8 : 5).

"In the day of my trouble I sought God with my hands lifted up to Him in the night, and I was not deceived. My soul remembered God, and was delighted, and was exercised, and my spirit swooned away. . . . And I said,

* "God cannot be remembered." If taken completely literally, just as it stands, without qualification, this statement would be false. We can have a valid conceptual knowledge of God. This knowledge can be stored in the memory and called back to mind. But the aphorism derives its point from the fact that there is another knowledge of God which goes beyond concepts, which passes through concepts to attain Him in the mysterious actuality of His presence, grasped in some sort, in an "experience." Even these experiences of God remain deeply engraved in the memory: but when we remember them, their actuality is no longer present, but past. This actuality of God's presence is something that does not belong to the past or to the future but only to the present. It cannot be brought back by an effort of memory, any more than it can be elicited by the work of imagination. It is a "discovery," and each time the discovery is new.

Now I have begun: this is the change of the right hand of the Most High" (Psalm 76: 3, 4, 11).

We could not seek God unless He were seeking us. We may begin to seek Him in desolation, feeling nothing but His absence. But the mere fact that we seek Him proves that we have already found Him. For if we continue in our prayer, we "remember" Him, that is to say, we become conscious, once again, of who He really is. And we see that He has found us. When this consciousness is the work of grace, it is always fresh and new. It is more than the recovery of a past experience. It is a new experience, and it makes us new men.

This newness is the "delight" and the "exercise" which are the living evidence of contact with the Spirit of the Lord. It makes us "swoon" in our spirit in a passage from death to life. Thus our eyes are opened. We see all things in a new light. And we realize that this is a new beginning, a change that could only be brought about by the intervention of His Spirit in our lives—the "change of the right hand of the Most High."

2. My Lord, You have heard the cry of my heart because it was You who cried out within my heart.

Forgive me for having tried to evoke Your presence in my own silence: it is You who must create me within Your own silence! Only this newness can save me from idolatry!

You are not found in the Temple merely by the expulsion of the money changers.

You are not found on the mountain every time there is a cloud. The earth swallowed those who offered incense without having been found, and called, and known by You.

3. If I find Him with great ease, perhaps He is not my God.

If I cannot hope to find Him at all, is He my God?

If I find Him wherever I wish, have I found Him?

If He can find me whenever He wishes, and tells me who He is and who I am, and if I then know that He whom I could not find has found me: then I know He is the Lord, my God: He has touched me with the finger that made me out of nothing.

4. A current of useless interior activity constantly surrounds and defends an illusion.

I cannot find God unless I renounce this useless activity, and I cannot renounce this activity unless I let go of the illusion it defends. And I cannot get rid of an illusion unless I recognize it for an illusion.

Interior silence is, therefore, not so much a negation, an absence of noise and of movement, as the positive rest of the mind in truth.

Man's intelligence, however we may misuse it, is far too keen and too sure to rest for long in error. It may embrace a lie and cling to it stubbornly, believing it to be true: but it cannot find true rest in falsehood. The mind that is in love with error wears itself out with anxiety, lest its error be discovered for what it is. But the man who loves truth can already find rest in the acknowledgment of his mistakes, for that is the beginning of truth.

The first step toward finding God, who is Truth, is to discover the truth about myself: and if I have been in error, this first step to truth is the discovery of my error. A false and illusory "experience" of what appears to be God's action in the soul may bring with it, for a moment, a kind of interior silence: the silence of a soul that rests in an illusion. But this silence is quickly disturbed by a deep undercurrent of unrest and noise. The tension of a soul trying to hold itself in silence, when it has no truth to appease it with a superior silence, is louder than the noise of big cities and more disturbing than the movement of an army.

5. The god of the philosophers lives in the mind that knows him, receives life by the fact that he is known, lives as long as he is known, and dies when he is denied. But the True God (whom the philosophers can truly find through their abstractions if they remember their vocation to pass beyond abstractions) gives life to the mind that is known by Him. The Living God, by the touch of His mercy in the depths of the soul that is "known" to His mercy, awakens the knowledge of His presence in that soul so that it not only knows Him but, at the same time, loves Him, seeing that it lives in Him. Therefore Jesus said: "The God of Abraham, the God of Isaac and the God of Jacob is not the God of the dead but of the living" (Matthew 22 : 32). So true is it that the Lord is the "Living God" that all those whose God He is will live forever, because He is their God. Such was the argument that Jesus gave to the Sadducees, who did not believe in the resurrection of the dead. If God was the "God of Abraham" then Abraham must rise from the dead: no one who has the Living God for his Lord can stay dead. He is our God only if we belong entirely to Him. To belong entirely to life is to have passed from death to life.

"For by a man came death, and by a man the resurrection of the dead. As in Adam all die, so in Christ all shall be made alive. . . . And the dead shall rise again incorruptible:

and we shall be changed. . . . And when this mortal hath put on immortality, then shall come to pass the saying that is written: 'Death is swallowed up in victory'" (I Corinthians 15 : 21–22, 52, 54).

We cannot find Him who is Almighty unless we are taken entirely out of our own weakness. But we must first find out our own nothingness before we can pass beyond it: and this is impossible as long as we believe in the illusion of our own power.

6. The monastery is the House of God and all who live there are close to Him. Yet it is possible to live close to Him and in His own house without ever finding Him. Why is this? Because we continue to seek ourselves rather than God, to live for ourselves rather than for Him. Then the monastery becomes our house rather than His, and He hides Himself from us. Standing in the way of His light, we gaze in perplexity upon our own shadow. "Surely," we say, "this has nothing of God in it, for it is a shadow." True! And yet the shadow is cast by His light, and bears witness indirectly to His presence. It is there to remind us that we can turn to Him whenever we cease to love darkness rather than the light.

Yet we fail to turn to Him because we forget that He must come to us as a saviour, without whom we are helpless. We seek Him as if He could not do without our sacrifices, as if He needed to be entertained by our affection and flattered by our praise.

We cannot find Him unless we know we need Him. We forget this need when we take a self-sufficient pleasure in our own good works. The poor and helpless are the first to find Him, who came to seek and to save that which was lost.

7. The Lord is my rock and my fortress and He dwells in the midst of His people.

Come, let us enter the House of the Almighty and stand to praise Him.

Let us sleep like eagles in the cliff, let us rest in the power of the Lord our God!

Let us hide ourselves in the great mountain of His might, who dwells concealed in the midst of a forsaken people.

Even His thunder is the refuge of the poor!

14

The Wind Blows Where It Pleases

GOD, who is everywhere, never leaves us. Yet He seems sometimes to be present, sometimes absent. If we do not know Him well, we do not realize that He may be more present to us when He is absent than when He is present.

There are two absences of God. One is an absence that condemns us, the other an absence that sanctifies us.

In the absence that is condemnation, God "knows us not" because we have put some other god in His place, and refuse to be known by Him. In the absence that sanctifies, God empties the soul of every image that might become an idol and of every concern that might stand between our face and His Face.

In the first absence, He is present, but His presence is denied by the presence of an idol. God is present to the enemy we have placed between ourselves and Him in mortal sin.

In the second absence He is present, and His presence is affirmed and adored by the absence of everything else. He is closer to us than we are to ourselves, although we do not see Him.

Whoever seeks to catch Him and hold Him loses Him. He is like the wind that blows where it pleases. You who love Him must love Him as arriving from where you do not know and as going where you do not know. Your spirit must seek to be as clean and as free as His own Spirit, in order to follow Him wherever He goes. Who are we to call ourselves either clean or free, unless He makes us so?

If He should teach us how to follow Him into the wilderness of His own freedom, we will no longer know where we are, because we are with Him who is everywhere and nowhere at the same time.

Those who love only His apparent presence cannot follow the Lord wherever He goes. They do not love Him perfectly if they do not allow Him to be absent. They do not respect His liberty to do as He pleases. They think their prayers have made them able to command Him, and to subject His will to their own. They live on the level of magic rather than on the level of religion.

Only those men are never separated from the Lord who never question His right to separate Himself from them. They never lose Him because they always realize they never deserve to find Him, and that in spite of their unworthiness they have already found Him.

For He has first found them, and will not let them go.

2. God approaches our minds by receding from them.

We can never fully know Him if we think of Him as an object of capture, to be fenced in by the enclosure of our own ideas.

We know Him better after our minds have let Him go.

The Lord travels in all directions at once.

The Lord arrives from all directions at once.

Wherever we are, we find that He has just departed.

Wherever we go, we discover that He has just arrived before us.

Our rest can be neither in the beginning of this pursuit,

nor in the pursuit itself, nor in its apparent end. For the true end, which is Heaven, is an end without end. It is a totally new dimension, in which we come to rest in the secret that He must arrive at the moment of His departure; His arrival is at every moment and His departure is not fixed in time.

3. Every man becomes the image of the god he adores.

He whose worship is directed to a dead thing becomes a dead thing.

He who loves corruption rots.

He who loves a shadow becomes, himself, a shadow.

He who loves things that must perish lives in dread of their perishing.

The contemplative also, who seeks to keep God prisoner in his heart, becomes a prisoner within the narrow limits of his own heart, so that the Lord evades him and leaves him in his imprisonment, his confinement, and his dead recollection.

The man who leaves the Lord the freedom of the Lord adores the Lord in His freedom and receives the liberty of the sons of God.

This man loves like God and is carried away, the captive of the Lord's invisible freedom.

A god who remains immobile within the focus of my own vision is hardly even a trace of the True God's passing.

4. What does it mean to know You, O my God?

There are souls who grow sick and tremble at the thought of giving You some insufficient name!

I wake up in the night and sweat with dread at the thought that I have dared to speak of You as "pure Being."

When Moses saw the bush in flames, burning in the desert but not consumed, You did not answer his question with a definition. You said, "*I am.*" What could be the effect of such an answer? It made the very dust of the earth instantly holy, so that Moses threw away his shoes (symbols of his senses and of his body) lest there remain any image in the way between Your sanctity and his adoration.

You are the strong God, the Holy, the Just One, strong and shy in Your vast mercy, hidden from us in Your freedom, giving us Your love without restraint, in order that, receiving all from You we may know that You alone are holy.

How shall we begin to know You who are if we do not begin ourselves to be something of what You are?

How can we begin to know You who are good unless we let You do us good?

How can we escape the knowledge of You who are good since no one can prevent You from doing good to us?

To "be" and to "be good" are things familiar to us. For we are made in Your image, with a being that is good because it is Your gift. But the being and the goodness we know fall so far short of You that they deceive us if we apply them to You as we know them in ourselves.

Therefore, they do not tell us, as they should, that You are holy.

5. The wise man has struggled to find You in his wisdom, and he has failed. The just man has striven to grasp You in his own justice, and he has gone astray.

But the sinner, suddenly struck by the lightning of mercy that ought to have been justice, falls down in adoration of Your holiness: for he had seen what kings desired to see and never saw, what prophets foretold and never gazed upon, what the men of ancient times grew weary of expecting when they died. He has seen that Your love is so infinitely good that it cannot be the object of a human bargain. True, there are two testaments, two bargains. But both of them are only promises that You would freely give us what we could never deserve: that You would manifest Your holiness to us by showing us Your mercy and Your liberality and Your infinite freedom.

"Is it not lawful to me," says the Lord, "to do what I will?" (Matthew 20:15). The supreme characteristic of His love is its infinite freedom. It cannot be compelled to respond to the laws of any desire, that is of any necessity. It is without limitation because it is without need. Being without need, His love seeks out the needy, not in order to give them a little but in order to give them all.

His love cannot be at peace in a soul that is content with a little. For to be content with a little is to will to continue in need.

It is not God's will that we should remain in need. He would fulfil all our needs by delivering us from all possessions and giving us Himself in exchange.

If we would belong to His love, we must remain always empty of everything else, not in order to be in need, but precisely because possessions make us needy.

6. Each true child of God is mild, perfect, docile, and alone. His consciousness springs up, in the Spirit of the Lord, at the precise point where he feels himself to be held in being by a pure gift, an act of love, a divine command.

The freedom of God's gift of life calls for the response of our own freedom—an act of obedience, hidden in the secrecy of our deepest being. We find the Lord when we find His gift of life in the depths of ourselves. We are fully alive in Him when the deepest roots of life become conscious that they live in Him. From this consent to exist in dependence upon His gift and upon His freedom springs the interior life.

7. Let the command of His love be felt at the roots of my existence.

Then let me understand that I do not consent in order to exist, but that I exist in order to consent.

This is the living source of virtuous action: for all our good acts are acts of consent to the indications of His mercy and the movements of His grace.

From this we can come to perfection: to the love which consents in all things, seeks nothing but to respond, by goodness, to Goodness, and by love, to Love. Such love suffers all things and is equally happy in action and inaction, in existence and in dissolution.

Let us not only exist, but *obey* in our existing.

From this fundamental obedience, which is a fundamental gift, and the fit return of His gift, all other acts of obedience spring up into life everlasting.

For the full fruitfulness of spiritual life begins in gratitude for life, in the consent to live, and in the greater gratitude that seeks to be dissolved and to be with Christ.

15

The Inward Solitude

CHARITY is a love for God which respects the need that other men have for Him. Therefore, charity alone can give us the power and the delicacy to love others without defiling their loneliness which is their need and their salvation.

2. Do not stress too much the fact that love seeks to penetrate the intimate secrets of the beloved. Those who are too fond of this idea fall short of true love, because they violate the solitude of those they love, instead of respecting it.

True love penetrates the secrets and the solitude of the beloved by allowing him to keep his secrets to himself and to remain in his own solitude.

3. Secrecy and solitude are values that belong to the very essence of personality.

A person is a person in so far as he has a secret and is a solitude of his own that cannot be communicated to anyone else. If I love a person, I will love that which most makes him a person: the secrecy, the hiddenness, the solitude of his own individual being, which God alone can penetrate and understand.

A love that breaks into the spiritual privacy of another

in order to lay open all his secrets and besiege his solitude with importunity does not love him: it seeks to destroy what is best in him and what is most intimately his.

4. Compassion and respect enable us to know the solitude of another by finding him in the intimacy of our own interior solitude. It discovers his secrets in our own secrets. Instead of consuming him with indiscretion, and thus frustrating all our own desires to show our love for him, if we respect the secrecy of his own interior loneliness, we are united with him in a friendship that makes us both grow in likeness to one another and to God. If I respect my brother's solitude, I will know his solitude by the reflection that it casts, through charity, upon the solitude of my own soul.

This respect for the deepest values hidden in another's personality is more than an obligation of charity. It is a debt we owe in justice to those who, like ourselves, are created in the image of God.

Our failure to respect the intimate spiritual privacy of other persons reflects a secret contempt for God Himself. It springs from the crass pride of fallen man, who wants to prove himself a god by prying into everything that is not his own business. The tree of the knowledge of good and evil gave our first parents a taste for knowing things outside of God, in a way in which they are not known truly, instead of knowing them in Him, in whom alone we are able to find them and know them and love them as they are. Original justice gave our souls the power to love well: to increase our own heritage of life by loving others for their own good. Original sin gave our souls the power to love destructively: to ruin the object of our love by consuming it, with no other profit to ourselves than the increase of our own interior famine.

We ruin others and ourselves together not by entering into the sanctuary of their inner being—for no one can enter there except their Creator—but by drawing them out of that sanctuary and teaching them to live as we live: centred upon themselves.

5. If a man does not know the value of his own loneliness, how can he respect another's solitude?

It is at once our loneliness and our dignity to have an incommunicable personality that is ours, ours alone and no one else's, and will be so forever.

When human society fulfils its true function the persons who form it grow more and more in their individual freedom and personal integrity. And the more each individual develops and discovers the secret resources of his own incommunicable personality, the more he can contribute to the life and the weal of the whole. Solitude is as necessary for society as silence is for language and air for the lungs and food for the body.

A community that seeks to invade or destroy the spiritual solitude of the individuals who compose it is condemning itself to death by spiritual asphyxiation.

6. If I cannot distinguish myself from the mass of other men, I will never be able to love and respect other men as I ought. If I do not separate myself from them enough to know what is mine and what is theirs, I will never discover what I have to give them, and never allow them the opportunity to give me what they ought. Only a *person* can pay debts and fulfil obligations, and if I am less than a

person I will never give others what they have a right to expect from me. If they are less than persons, they will not know what to expect from me. Nor will they ever discover that they have anything to give. We ought normally to educate one another by fulfilling one another's just needs. But in a society where personality is obscured and dissolved, men never learn to find themselves and, therefore, never learn how to love one another.

7. Solitude is so necessary both for society and for the individual that when society fails to provide sufficient solitude to develop the inner life of the persons who compose it, they rebel and seek false solitudes.

A false solitude is a point of vantage from which an individual, who has been denied the right to become a person, takes revenge on society by turning his individuality into a destructive weapon. True solitude is found in humility, which is infinitely rich. False solitude is the refuge of pride, and it is infinitely poor. The poverty of false solitude comes from an illusion which pretends, by adorning itself in things it can never possess, to distinguish one individual self from the mass of other men. True solitude is selfless. Therefore, it is rich in silence and charity and peace. It finds in itself seemingly inexhaustible resources of good to bestow on other people. False solitude is self-centred. And because it finds nothing in its own centre, it seeks to draw all things into itself. But everything it touches becomes infected with its own nothingness, and falls apart. True solitude cleans the soul, lays it wide open to the four winds of generosity. False solitude locks the door against all men and pores over its own private accumulation of rubbish.

Both solitudes seek to distinguish the individual from the crowd. True solitude succeeds in this, false solitude fails. True solitude separates one man from the rest in order that he may freely develop the good that is his own, and then fulfil his true destiny by putting himself at the service of everyone else. False solitude separates a man from his brothers in such a way that he can no longer effectively give them anything or receive anything from them in his own spirit. It establishes him in a state of indigence, misery, blindness, torment, and despair. Maddened by his own insufficiency, the proud man shamelessly seizes upon satisfactions and possessions that are not due to him, that can never satisfy him, and that he will never really need. Because he has never learned to distinguish what is really his, he desperately seeks to possess what can never belong to him.

In reality the proud man has no respect for himself because he has never had an opportunity to find out if there is anything in him worthy of respect. Convinced that he is despicable, and desperately hoping to keep other men from finding it out, he seizes upon everything that belongs to them and hides himself behind it. The mere fact that a thing belongs to someone else makes it seem worthy of desire. But because he secretly hates everything that is his own, as soon as each new thing becomes his own it loses its value and becomes hateful to him. He must fill his solitude with more and more loot, more and more rapine, seizing things not because he wants them, but because he cannot stand the sight of what he has already obtained.

These, then, are the ones who isolate themselves above the mass of other men because they have never learned to love either themselves or other men. They hate others because they hate themselves, and their love of others is merely an expression of this solitary hatred.

The proud solitary is never more dangerous than when he appears to be social. Having no true solitude and, therefore, no spiritual energy of his own, he desperately needs other men. But he needs them in order to consume them, as if in consuming them he could fill the void in his own spirit and make himself the person he feels he ought to be.

When the Lord, in His justice, wills to manifest and punish the sins of a society that ignores the natural law, He allows it to fall into the hands of men like this. The proud solitary is the ideal dictator, turning the whole world from peace to war, carrying out the work of destruction, opening the mouths of ruin from city to city, that these may declare the emptiness and degradation of men without God.

The perfect expression of a society that has lost all sense of the value of personal solitude is a state forced to live as a refugee in its own ruins, a mob without roofs to cover it, a herd without a barn.

8. True solitude is the solitude of charity, which "seeketh not her own."* It is ashamed to have anything that is not due to it. It seeks poverty, and desires to give away all that it does not need. It seems to feel distaste for created things: but its distaste is not for them. It cannot hate them, for it cannot even hate itself. Because it loves them, it knows it cannot own them, since they belong to God. Charity desires that He alone should possess them and receive from them the glory which is His due.

* "*Caritas . . . non quaerit quae sua sunt*" (I Corinthians 13 : 5).

Our solitude may be fundamentally true, but still imperfect. In that event, it is mixed with pride. It is a disturbing mixture of hatred with love. One of the secrets of spiritual perfection is to realize that we have this mixture in ourselves, and to be able to distinguish one from the other. For the temptation of those who seek perfection is to mistake hatred for love, and to place their perfection in the solitude which distinguishes itself from other men by hating them and which at the same time loves and hates the good things that are theirs.

The asceticism of the false solitary is always double-dealing. It pretends to love others, but it hates them. It pretends to hate created things, and it loves them. And by loving them in the wrong way, it only succeeds in hating them.

Therefore, as long as our solitude is imperfect it will be tainted with bitterness and disgust, because it will exhaust us in continual conflict. The disgust is unavoidable. The bitterness, which should not be, is, nevertheless, there. Both must be used for our purification. They must teach us to distinguish what is truly bitter from what is truly sweet, and not permit us to find a poisoned sweetness in self-hatred and a poisoned bitterness in the love of others.

The true solitary must recognize that he is obliged to love other men and even all things created by God: that this obligation is not a painful and unpleasant duty, and that it was never supposed to be bitter. He must accept the sweetness of love without complaint, and not hate himself because his love may be, at first, a little inordinate. He must suffer without bitterness in order to learn to love as he ought. He must not fear that love will destroy his solitude. Love *is* his solitude.

9. Our solitude will be imperfect as long as it is tainted with restlessness and *accedia*. For the vice of *accedia* makes us hate what is good and shrink from the only virtues that can save us. Pure interior solitude does not shrink from the good things of life or from the company of other men, because it no longer seeks to possess them for their own sake. No longer desiring them, it no longer fears to love them. Free from fear, it is free of bitterness. Purified of bitterness, the soul can safely remain alone.

Indeed, the soul that does not seek to dress itself in possessions and to revel in purchased or stolen satisfactions will often be left completely alone by other men. The true solitary does not have to run away from others: they cease to notice him, because he does not share their love for an illusion. The soul that is truly solitary becomes perfectly colourless and ceases to excite either the love or the hatred of others by reason of its solitude. The true solitary can, no doubt, become a hated and a hunted person: but not by reason of anything that is in himself. He will only be hated if he has a divine work to do in the world. For his work will bring him into conflict with the world. His solitude, as such, creates no such conflict. Solitude brings persecution only when it takes the form of a "mission," and then there is something much more in it than solitude. For when the solitary finds that his solitude has taken on the character of a mission, he discovers that he has become a force that reacts on the very heart of the society in which he lives, a power that disturbs and impedes and accuses the forces of selfishness and pride, reminding others of their own need for solitude and for charity and for peace with God.

10. Pure interior solitude is found in the virtue of hope. Hope takes us entirely out of this world while we remain bodily in the midst of it. Our minds retain their clear view of what is good in creatures. Our wills remain chaste and solitary in the midst of all created beauty, not wounded in an isolation that is prudish and ashamed, but lifted up to Heaven by a humility that hope has divested of all bitterness, all consolation, and all fear.

Thus we are both in time and out of it. We are poor, possessing all things. Having nothing of our own left to rely on, we have nothing to lose and nothing to fear. Everything is locked away for our sure possession, beyond our reach, in Heaven. We live where our souls desire to be, and our bodies no longer matter very much. We are buried in Christ, our life is hidden with Christ in God and we know the meaning of His freedom.

This is true solitude, about which there are no disputes and no questions. The soul that has thus found itself gravitates toward the desert but does not object to remaining in the city, because it is everywhere alone.

16

Silence

THE rain ceases, and a bird's clear song suddenly announces the difference between Heaven and hell.

2. God our Creator and Saviour has given us a language in which He can be talked about, since faith cometh by hearing and our tongues are the keys that open Heaven to others.

But when the Lord comes as a Bridegroom there remains nothing to be said except that He is coming, and that we must go out to meet Him. *Ecce Sponsus venit! Exite obviam ei!**

After that we go forth to find Him in solitude. There we communicate with Him alone, without words, without discursive thoughts, in the silence of our whole being.

When what we say is meant for no one else but Him, it can hardly be said in language. What is not meant to be related is not even experienced on a level that can be clearly analysed. We know that it must not be told, because it cannot.

But before we come to that which is unspeakable and unthinkable, the spirit hovers on the frontiers of language, wondering whether or not to stay on its own side of the

* "Behold the Bridegroom cometh, go ye forth to meet Him" (Matthew 25 : 6).

border, in order to have something to bring back to other men. This is the test of those who wish to cross the frontier. If they are not ready to leave their own ideas and their own words behind them, they cannot travel farther.

3. Do not desire chiefly to be cherished and consoled by God; desire above all to love Him.

Do not anxiously desire to have others find consolation in God, but rather help them to love God.

Do not seek consolation in talking about God, but speak of Him in order that He may be glorified.

If you truly love Him, nothing can console you but His glory. And if you seek His glory before everything else, then you will also be humble enough to receive consolation from His hand: accepting it chiefly because, in showing His mercy to us, He is glorified in our souls.

If you seek His glory before everything else, you will know that the best way to console another man is to show him how to love God. There is no true peace in anything else.

If you wish your words about Him to mean something, they must be charged with zeal for His glory. For if your hearers realize that you are speaking only to please yourself, they will accuse your God of being nothing more than a shadow. If you love His glory, you will seek this transcendence—and this is sought in silence.

Let us, then, not seek comfort in the assurance that we are good, but only in the certainty that He alone is holy, He alone is good.

It is not seldom that our silence and our prayers do more to bring people to the knowledge of God than all our words about Him. The mere fact that you wish to give

God glory by talking about Him is no proof that your speech will give Him glory. What if He should prefer you to be silent? Have you never heard that silence gives Him glory?

4. If you go into solitude with a silent tongue, the silence of mute beings will share with you their rest.

But if you go into solitude with a silent heart, the silence of creation will speak louder than the tongues of men or angels.

5. The silence of the tongue and of the imagination dissolves the barrier between ourselves and the peace of things that exist only for God and not for themselves. But the silence of all inordinate desire dissolves the barrier between ourselves and God. Then we come to live in Him alone.

Then mute beings no longer speak to us merely with their own silence. It is the Lord who speaks to us, with a far deeper silence, hidden in the midst of our own selves.

6. Those who love their own noise are impatient of everything else. They constantly defile the silence of the forests and the mountains and the sea. They bore through silent nature in every direction with their machines, for fear that the calm world might accuse them of their own emptiness. The urgency of their swift movement seems to ignore the tranquillity of nature by pretending to have a purpose. The loud plane seems for a moment to deny the reality of the clouds and of the sky, by its direction, its noise, and its pretended strength. The silence of the sky remains when the plane has gone. The tranquillity of the clouds will remain when the plane has fallen apart. It is the silence of the world

that is real. Our noise, our business, our purposes, and all our fatuous statements about our purposes, our business, and our noise: these are the illusion.

God is present, and His thought is alive and awake in the fullness and depth and breadth of all the silences of the world. The Lord is watching in the almond trees, over the fulfilment of His words (Jeremias 1 : 11).

Whether the plane pass by tonight or tomorrow, whether there be cars on the winding road or no cars, whether men speak in the field, whether there be a radio in the house or not, the tree brings forth her blossoms in silence.

Whether the house be empty or full of children, whether the men go off to town or work with tractors in the fields, whether the liner enters the harbour full of tourists or full of soldiers, the almond tree brings forth her fruit in silence.

7. There are some men for whom a tree has no reality until they think of cutting it down, for whom an animal has no value until it enters the slaughterhouse, men who never look at anything until they decide to abuse it and who never even notice what they do not want to destroy. These men can hardly know the silence of love: for their love is the absorption of another person's silence into their own noise. And because they do not know the silence of love, they cannot know the silence of God, who is Charity, who cannot destroy what He loves, who is bound, by His own law of Charity, to give life to all those whom He draws into His own silence.

8. Silence does not exist in our lives merely for its own sake. It is ordered to something else. Silence is the mother of

speech. A lifetime of silence is ordered to an ultimate declaration, which can be put into words, a declaration of all we have lived for.

Life and death, words and silence, are given us because of Christ. In Christ we die to the flesh and live to the spirit. In Him we die to illusion and live to truth. We speak to confess Him, and we are silent in order to meditate on Him and enter deeper into His silence, which is at once the silence of death and of eternal life—the silence of Good Friday night and the peace of Easter morning.

9. We receive Christ's silence into our hearts when first we speak from our heart the word of faith. We work out our salvation in silence and in hope. Silence is the strength of our interior life. Silence enters into the very core of our moral being, so that if we have no silence we have no morality. Silence enters mysteriously into the composition of all the virtues, and silence preserves them from corruption.

By the "silence" of virtue I mean the charity which must give each virtue a supernatural life and which is "silent" because it is rooted in God. Without this silence, our virtues are sound only, only an outward noise, a manifestation of nothing: the thing that virtues manifest is their own interior charity, which has a "silence" of its own. And in this silence hides a Person: Christ, Himself hidden, as He is spoken, in the silence of the Father.

10. If we fill our lives with silence, then we live in hope, and Christ lives in us and gives our virtues much substance. Then, when the time comes, we confess Him openly

before men, and our confession has much meaning because it is rooted in deep silence. It awakens the silence of Christ in the hearts of those who hear us, so that they themselves fall silent and begin to wonder and to listen. For they have begun to discover their true selves.

If our life is poured out in useless words we will never hear anything in the depths of our hearts, where Christ lives and speaks in silence. We will never be anything, and in the end, when the time comes for us to declare who and what we are, we shall be found speechless at the moment of the crucial decision: for we shall have said everything and exhausted ourselves in speech before we had anything to say.

11. There must be a time of day when the man who makes plans forgets his plans, and acts as if he had no plans at all.

There must be a time of day when the man who has to speak falls very silent. And his mind forms no more propositions, and he asks himself: Did they have a meaning?

There must be a time when the man of prayer goes to pray as if it were the first time in his life he had ever prayed; when the man of resolutions puts his resolutions aside as if they had all been broken, and he learns a different wisdom: distinguishing the sun from the moon, the stars from the darkness, the sea from the dry land, and the night sky from the shoulder of a hill.

12. In silence, we learn to make distinctions. Those who fly silence, fly also from distinctions. They do not want to see too clearly. They prefer confusion.

A man who loves God necessarily loves silence also,

because he fears to lose his sense of discernment. He fears the noise that takes the sharp edge off every experience of reality. He avoids the unending movement that blurs all beings together into a crowd of undistinguishable things.

The saint is indifferent in his desires, but by no means indifferent in his attitudes toward different aspects of reality.

13. Here lies a dead man who made an idol of indifference.

His prayer did not enkindle, it extinguished his flame.

His silence listened to nothing and, therefore, heard nothing, and had nothing to say.

Let the swallows come and build their nests in his history and teach their young to fly about in the desert which he made of his soul, and thus he will not remain unprofitable forever.

14. Life is not to be regarded as an uninterrupted flow of words which is finally silenced by death. Its rhythm develops in silence, comes to the surface in moments of necessary expression, returns to deeper silence, culminates in a final declaration, then ascends quietly into the silence of Heaven which resounds with unending praise.

Those who do not know there is another life after this one, or who cannot bring themselves to live in time as if they were meant to spend their eternity in God, resist the fruitful silence of their own being by continual noise. Even when their own tongues are still, their minds chatter without end and without meaning, or they plunge themselves into the protective noise of machines, traffic, or radios. When their own noise is momentarily exhausted, they rest in the noise of other men.

How tragic it is that they who have nothing to express are continually expressing themselves, like nervous gunners, firing burst after burst of ammunition into the dark, where there is no enemy. The reason for their talk is: death. Death is the enemy who seems to confront them at every moment in the deep darkness and silence of their own being. So they keep shouting at death. They confound their lives with noise. They stun their own ears with meaningless words, never discovering that their hearts are rooted in a silence that is not death but life. They chatter themselves to death, fearing life as if it were death.

15. Our whole life should be a meditation of our last and most important decision: the choice between life and death.

We must all die. But the dispositions with which we face death make of our death a choice either of death or of life.

If, during our life we have chosen life, then in death we will pass from death to life. Life is a spiritual thing, and spiritual things are silent. If the spirit that kept the flame of physical life burning in our bodies took care to nourish itself with the oil that is found only in the silence of God's charity, then when the body dies, the spirit itself goes on burning the same oil, with its own flame. But if the spirit has burned all along with the base oils of passion or egoism or pride, then when death comes the flame of the spirit goes out with the light of the body because there is no more oil in the lamp.

We must learn during our lifetime to trim our lamps and fill them with charity in silence, sometimes speaking and confessing the glory of God in order to increase our charity by increasing the charity of others, and teaching them also the ways of peace and of silence.

16. If, at the moment of our death, death comes to us as an unwelcome stranger, it will be because Christ also has always been to us an unwelcome stranger. For when death comes, Christ comes also, bringing us the everlasting life which He has bought for us by His own death. Those who love true life, therefore, frequently think about their death. Their life is full of a silence that is an anticipated victory over death. Silence, indeed, makes death our servant and even our friend. Thoughts and prayers that grow up out of the silent thought of death are like trees growing where there is water. They are strong thoughts, that overcome the fear of misfortune because they have overcome passion and desire. They turn the face of our soul, in constant desire, toward the face of Christ.

17. If I say that a whole lifetime of silence is ordered to a final utterance, I do not mean that we must all contrive to die with pious speeches on our lips. It is not necessary that our last words should have some special or dramatic significance worthy of being written down. Every good death, every death that hands us over from the uncertainties of this world to the unfailing peace and silence of the love of Christ, is itself an utterance and a conclusion. It says, either in words or without them, that it is good for life to come to its appointed end, for the body to return to dust and for the spirit to ascend to the Father, through the mercy of Our Lord Jesus Christ.

A silent death may speak with more eloquent peace than a death punctuated by vivid expressions. A lonely death, a tragic death, may yet have more to say of the peace and mercy of Christ than many another comfortable death.

For the eloquence of death is the eloquence of human poverty coming face to face with the riches of divine mercy. The more we are aware that our poverty is supremely great, the greater will be the meaning of our death: and the greater its poverty. For the saints are those who wanted to be poorest in life, and who, above all else, exulted in the supreme poverty of death.